# The Invisible Universe Revealed

Gerrit L. Verschuur

# The Invisible Universe Revealed

## The Story of Radio Astronomy

With 108 Figures

Springer-Verlag
New York Berlin Heidelberg
London Paris Tokyo

Gerrit L. Verschuur
National Radio Astronomy Observatory
Charlottesville, Virginia 22903, USA

The cover photo of the National Radio Astronomy Observatory's Very Large Array (VLA) was taken by moonlight (f/4; approximately 20-min exposure) by Geoffrey R. Chester, Production Coordinator, Albert Einstein Planetarium, Smithsonian Institution. Mr. Chester used a Nikon FM2 camera with a Nikkor 16-mm fisheye lens; the film was Ektachrome 200.

Library of Congress Cataloging in Publication Data
Verschuur, Gerrit L., 1937–
    The invisible universe revealed.
    "Completely rewritten second edition of The
invisible universe . . . originally published in
1974"—Pref.
    Bibliography: p.
    Includes index.
    1. Radio astronomy.   I. Verschuur, Gerrit L.,
1937–    .   Invisible universe.   II. Title.
QB475.V47   1986     522'.682          86–20321

This book is a completely rewritten version of *The Invisible Universe: The Story of Radio Astronomy* originally published by Springer-Verlag in 1974.

Typeset by Arcata Graphics/Kingsport, Kingsport, Tennessee.
Printed and bound by Quinn-Woodbine, Woodbine, New Jersey.
Printed in the United States of America.

9 8 7 6 5 4 3 2 1

ISBN 0-387-96280-8 Springer-Verlag New York Berlin Heidelberg
ISBN 3-540-96280-8 Springer-Verlag Berlin Heidelberg New York

*To my wife Joan Schmelz and my son Carl
for their love and understanding*

# Preface

*The Invisible Universe Revealed: The Story of Radio Astronomy* is a completely rewritten version of *The Invisible Universe: The Story of Radio Astronomy* originally published in 1974. The new title reflects on the remarkable improvements made in the radio astronomer's ability to *see* the radio universe more clearly.

The book is nontechnical and aimed at the interested lay person as well as students in high schools and colleges. It may be effectively used as supplementary reading at college level.

Historical material has been kept to a minimum, not because it isn't important or interesting, but because several books have recently done the history justice. I have also mentioned few people by name, a decision made in the interests of telling a story. So many people were involved in making the important discoveries that there are too many names to mention, a sign that this science is maturing rapidly. I apologize to those who may feel slighted.

This book was made possible by the kindness of Drs. Hein Hvatum and Paul Vanden Bout in welcoming me to the National Radio Astronomy Observatory while I worked on the project. The staff were wonderfully cooperative and made my visit a genuine thrill. It gives me special pleasure to thank Drs. Ken Kellermann, Harvey Liszt, Steve Reynolds, Dan Stinebring, and Robert Hjellming for helpful comments on early versions of some of these chapters and Dr. Richard Thompson for the title idea. Dr. Bob Havlen's general support is highly valued. Joan Schmelz, Nancy Wiener, Carl Verschuur, and Deborah Ryan ploughed through parts or all of the original manuscript, and their patience and valuable comments are deeply appreciated.

This book could not have been created without the cooperation of the dozens of radio astronomers all over the world who provided the original illustrations as well as those who sent illustrations which, unfortunately, could not be included. Many of the illustrations originally appeared in *The Astrophysical Journal*.

I especially wish to thank Von Del Chamberlain, and Drs. René Walterbos, Ernie Seaquist, Doug Milne, and Dennis Downes. The extra assistance of Peggy Weems, Ron Monk, Pat Smiley, and Elaine Gardner Ollis in matters regarding the preparation of illustrations is highly valued. For special on-site help in the production of some of the VLA images I am grateful to Drs. Arnold Rots, Ron Ekers, Rick Perley, Pat Crane, and Carl Bignell.

The project would have been impossible without discussions regarding its contents, and my education benefited from talks with Drs. Barry Turner, Jim Condon, Dave Hogg, Dan Stinebring, Alice Hine, Butler Burton, Dick Tompson, John Findlay, Don Campbell, Bob Hjellming, and Chris O'Dea, as well as many radio astronomers whose brains I picked during seminars and conversations. My thanks also to the NRAO librarians Ellen Bouton and Mary Jo Hendricks whose excellent library allowed me to do the research relatively painlessly, and finally to Joan Schmelz for her enthusiastic support, encouragement, and help.

The Astronomical Society of the Pacific, in collaboration with the National Radio Astronomy Observatory and this author, has produced a slide set entitled ''The Radio Universe'' which consists of over 40 images, many of them color versions of the radiographs in this book. The set is ideal for teaching or planetarium use and is accompanied by a booklet describing the images. These slides can be effectively used in conjunction with *The Invisible Universe Revealed*. For information write to: The Astronomical Society of the Pacific, 1290 24th Avenue, San Francisco, CA 94122. For hobbyists interested in radio astronomy, The Society of Amateur Radio Astronomers (SARA) is an international nonprofit group that supports the efforts of amateurs in this discipline. For information write: SARA, 7605 Deland Avenue, Fort Pierce, FL 33451.

Many of the images in this book were produced and made available through the National Radio Astronomy Observatory, operated by Associated Universities, Inc. under contract with the National Science Foundation. This credit has been abbreviated to ''NRAO'' in the relevant captions.

Charlottesville, Virginia                                    Gerrit L. Verschuur

# Contents

## 9. Interstellar Hydrogen

## 10. Interstellar Molecules

## Part IV. Stellar-Type Radio Sources

## 11. Pulsars

## 12. The Radio Sun and Planets

## 13. The Galactic Superstars

## Part V. The Universe and Life

## 14. Beyond the Quasars—Radio Cosmology

## 15. On the Search for Extraterrestrial Intelligence

## Part VI. Radio Astronomy Review: Past, Present, and Future

## 16. Musings on the Evolution of a Science

## 17. Radio Telescopes —The Present

## 18. The Future

## Appendix

## Further Reading

## Index

# Part I

## Introduction

# 1

# The Adventure of Radio Astronomy

## What Is It?

Radio astronomy is the study of radio waves from space. Many objects in the universe, including certain stars, galaxies, and nebulae, as well as a wide variety of peculiar, fascinating, and often mysterious objects, emit radio waves through naturally occurring processes. Radio astronomers use large antennas, known as radio telescopes (Figure 1.1), together with very sensitive radio receivers to detect the presence of the waves, and our book is about the findings of this young, twentieth-century science. (See Appendix for definitions.)

The first cosmic radio waves were discovered in the early 1930s by Karl Jansky, a physicist at the Bell Labs, in Holmdell, New Jersey, who was trying to find out why transatlantic radiotelephone circuits suffered from a variety of hissing and crackling sounds. He soon identified thunderstorm activity as the source of clicks and bangs, but he also heard a faint and steady hiss which seemed to be reaching the earth from the direction of the Milky Way. This discovery marked the birth of radio astronomy.

At first Jansky's discovery went unrecognized by the astronomical community, and it was Grote Reber, an engineer in Wheaton, Illinois, who built the world's first dish-shaped radio telescope—in his own backyard! We have come full circle because such dishes are now found in backyards all across the country. Reber spent years studying the cosmic radio waves while astronomers still did not get involved. Those who heard about the new phenomenon knew too little about electronics to build the necessary equipment, and then the world was rocked by the Second World War. Radio astronomy was, effectively, "on hold" until 1946. In that year, astronomers and physicists began to take an active interest in the "newfangled wireless astronomy," especially because surplus radar equipment was available. The radar antennas, often in the shape of small dishes, and receivers were just what was needed to tune into the cosmic radio waves.

One of the earliest postwar discoveries was that specific regions of the sky seemed to emit more radio energy than their surroundings. These were given the generic name of *radio source*. Whenever a larger radio telescope or more

**FIGURE 1.1.** The 100-meter-diameter radio telescope in Bad Münstereifel-Effelsberg, near Bonn, West Germany. Radio waves from space reflect off the metal surface and are brought to a focus just below the structure supported above the center of the dish. There an antenna connected to a sensitive radio amplifier converts the radio waves into electrical impulses for further amplification and processing by computers in the main control room. (Max-Planck-Institut für Radioastronomie.)

sensitive radio receiver was used, more radio sources were discovered. Today tens of thousands of radio sources are known.

The accuracy with which the first radio sources were located in the sky was insufficient to allow optical astronomers to decide which of the hundreds of thousands of stars, galaxies, and nebulae whose images appeared in their photographs of the region in question were responsible for the radio emission. In order to make an optical identification, the astronomers required an accuracy of one arcminute (see Appendix, p. 250) or less. Such positional accuracy did not become commonly feasible until the mid-1960s, although by the late 1940s and early 1950s half a dozen of the strongest radio sources had been identified with optical objects. These included a couple of nebulae associated with the remains of exploded stars and several distant galaxies. One of those galaxies was in the constellation Cygnus, and in those days was detected as a small increase in radio energy from its direction. Today, with modern radio telescopes, this source is revealed in glorious detail, as shown in Figure 1.2, an image which signifies the extraordinary progress that has occurred in this science since its birth.

The list of radio sources now known includes stars, nebulae, galaxies, quasars, pulsars, and the sun and its planets, as well as amazing clouds of molecules between the stars, all of which generate radio waves. All of these types of objects will be discussed in our story. The study of the cosmic radio waves— where they come from, how they are produced, what sorts of astronomical objects are involved—is what radio astronomy is all about.

**FIGURE 1.2.** A dramatic radio image, or *radiograph,* of a section of the radio source Cygnus A, a radio galaxy located $6 \times 10^8$ light-years away. Details as small as a half-second of arc across, comparable to the clarity of vision obtained with the world's best optical telescopes, can be seen. The extent of the radio lobe is approximately $2 \times 10^5$ light-years across. This image illustrates the fabulous clarity with which radio astronomers now observe the otherwise invisible universe of radio sources. Vertical size is $30''$. (NRAO. Observers—R. A. Perley, J. J. Cowan, and J. W. Dreher.)

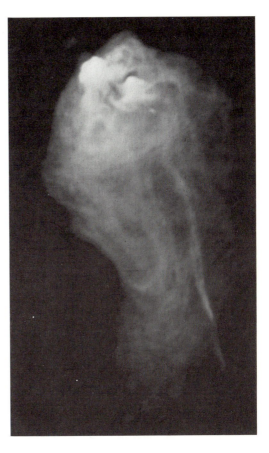

# The Exploration of the Radio Astronomical Unknown

Modern radio astronomy is one of the great adventures of the human spirit. Exploratory behavior, the primal urge to explore the unknown, is expressed in a deep human hunger for venturing into new worlds, a hunger that has been dramatically expressed in thousands of years of slow, systematic, and sometimes frightening journeys of exploration. These journeys, overland and across the seas and oceans, have carried people from their birthplaces to the most distant corners of the planet. Now humans have conquered terrestrial space and, like pollen on the wind, our species has moved from the caves of the earth to the craters of the moon. Our instinct drives us on, not just to the planets, but farther, into the universe beyond our senses. In that astronomical universe, profound mysteries have been uncovered, mysteries which challenge our imagination and our capacity for comprehension.

Radio waves from space are a form of energy which carries information about some of the most intriguing natural phenomena yet discovered by human beings. This information is in the air all about us—if we could only sense it in the right way. The cosmic radio whispers compete with the din produced by the radio energies generated from TV, radio, FM, radar, satellite, and radiotelephone channels. Cosmic radio waves from distant cosmic catastrophes, from the birth and death of stars and from the explosions of galaxies well beyond sight, are nearly lost in the man-made static. The radio waves from space in which we are bathed contain the secrets of interstellar gas clouds and carry messages from the remnants of the big bang which propelled our universe into existence. In order to focus in on the faint cosmic signals and avoid the unwanted stuff, the radio astronomer uses powerful radio telescopes located far from cities. The telescopes are equipped with sensitive receivers which amplify the received radio waves and feed them to computers, where they are converted into some visual form so that the data can be displayed, analyzed, interpreted, and hopefully understood.

The story of modern radio astronomy is a tale of the constant quest to express in clearer visual form the information carried by the radio waves. For this reason, radio astronomers are always inventing new ways to allow them to see the radio sources more clearly. The better we see the radio sources, the greater the likelihood that we will someday be able to understand their inner secrets.

Ever since Galileo first turned a telescope toward the heavens in 1609 A.D., centuries of technological innovation have afforded ever clearer views of astronomical objects in the far reaches of space. Larger and more sophisticated telescopes are always being designed and constructed. The Hubble Space Telescope, the mightiest optical telescope yet built, will allow astronomers to see the visible universe with fabulous new clarity. Modern technological marvels, which include the giant radio telescopes now revealing the radio universe to our gaze, have opened our imaginations to the universe beyond the senses in a way undreamed of even two decades ago.

## "Seeing" Radio Waves

Radio astronomers always talk about "seeing" radio sources or "seeing" radio waves. This is a figure of speech, because they do not literally "see" radio waves, nor do they even "listen to sounds" from space. Cosmic radio whispers are usually far too faint for the human ear to perceive, even after the radio signals have been amplified a million times. Modern computers and graphics technology allow the radio waves to be converted into electrical signals which can be combined to produce a photograph of what the radio source would look like if you had radio-sensitive eyes. Such an image is called a *radiograph*, and Figure 1.2 is an example. All forms of data produced by radio telescopes, whether radiographs or just sets of numbers, are examined visually, so astronomers talk as if they can "see" the objects of their study.

(The reader already familiar with some of the basic concepts of radio astronomy can readily move on to Chapter 2 now.)

## The Electromagnetic Spectrum

Electromagnetic radiation can be thought of as having a wavelike nature. Radio and light waves are both part of the *electromagnetic spectrum*, the name given to the range of waves which share common electrical and magnetic properties. All electromagnetic waves, including light, travel at a speed of 300,000 kilometers per second (km/s). The distance between the peaks of the waves is known as the *wavelength*. Radio waves have wavelengths that range from about one millimeter (1 mm) to hundreds of meters. "Microwaves" have wavelengths of around a few centimeters. The wavelength of light ranges from 40-millionths of a centimeter ($4 \times 10^{-5}$ cm), for violet light, to 70-millionths of a centimeter ($7 \times 10^{-5}$ cm) for the longer-wavelength red light. (See Notation, in the Appendix, for an explanation of scientific notation.) The colors of the rainbow fall between these two extremes. (The Appendix includes a discussion of the entire electromagnetic spectrum, which includes radio, infrared, light, ultraviolet, X-ray, and gamma-ray waves, and a discussion of the wavelength "windows" through which these waves pass on their way from space to the surface of the earth.) Only light and radio waves travel relatively unhindered through the atmosphere. In order to do astronomy at the other wavelengths, astronomers use satellites or balloon-borne telescopes in order to get above the obscuring layer of air.

## Radio Telescopes

The telescopes with which radio astronomers "see" the invisible universe are usually giant dish-shaped steel reflectors. These intercept radio waves over their entire area and reflect them to a concentrated focus. The dishes act as giant buckets to gather radio energy. A single-dish radio telescope, such as the 100-meter-diameter telescope near Bonn, West Germany, shown in Figure 1.1, is a large-scale version of the satellite dishes often found in rural backyards, on cable company premises, or on farms all over the USA. Radio telescopes are much larger and are equipped with extremely sensitive amplifiers so that they can detect faint radio whispers from very distant astronomical objects.

A single-dish radio telescope will collect all the radio energy coming from some small area in the heavens at any instant. If it should be pointed at a TV satellite it will be swamped by those signals, but the radio astronomer obviously avoids pointing in those directions! The area of the sky observed by the radio telescope at any given time defines the *beam*, and the angular extent of the beam is called the *beamwidth*, which, for a single dish, may be degrees or arcminutes (see Astronomical Coordinate Systems in the Appendix) across. In order to produce the equivalent of a photograph of a section of sky, the radio telescope has to be systematically "scanned," in the same way that a TV image is produced by scanning an electron beam across the TV screen. The intensity of received radio signals is recorded and the data combined to produce the radiograph, the visual image of what a particular direction in the sky looks like to the radio telescope. Figure 1.3 is a radiograph of the remarkable object known as 3C 75, which consists of four huge jets of matter spewed out from two bright cores of what may be colliding galaxies.

Note that the most intense radio source in a given constellation was originally given the name of the constellation followed by the letter A, e.g., Taurus A, Cygnus A (Figure 1.2), and so on. Today sources are often known by their

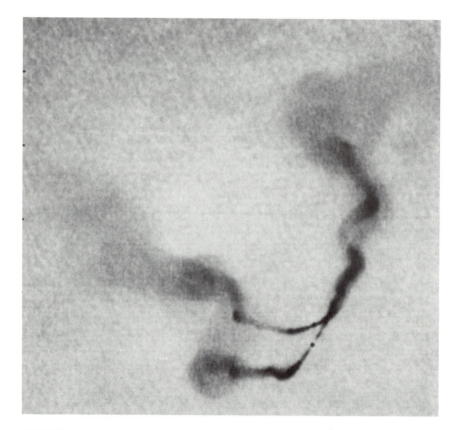

**FIGURE 1.3.** Radiograph of the spectacular radio source 3C 75, located in a cluster of galaxies $3 \times 10^8$ light-years away. This strange object contains two bright centers about $2 \times 10^4$ light-years apart, each the source of twin radio jets which appear twisted as they splay out into space. The entire radio-emitting object is about $10^6$ light-years across. Vertical size is $5''$. (NRAO. Observers—F. N. Owen, C. P. O'Dea, and M. Inoue.)

3C names, referring to the Third Cambridge Catalog of radio sources (e.g., 3C 75), or by names which refer to their positions on the sky.

## How Radio Waves from Space Are Generated

Cosmic radio waves are created in several ways, depending on the physical conditions in the radio-emitting object. All the processes involve the movement of electrons, in particular changes in their velocity during which they lose energy which can be radiated away as a radio wave. Radio energy is produced either by slow-moving electrons (traveling in the range of tens to hundreds of kilometers per second) within hot clouds of gas which surround very hot stars or by electrons which have been accelerated to travel close to the speed of light as the result of stellar or larger-scale explosions, which energize the particles. The two radio emission processes are known, respectively, as *thermal* and *nonthermal.*

Thermal radio emission is produced by electrons in a cloud of ionized matter (see Appendix, p. 244), where the temperature may be between 5000 and 20,000 degrees Kelvin. [Note: Astronomers refer to the temperature of an object in terms of degrees Kelvin (K). A degree Kelvin has the same magnitude as a degree centigrade, but the Kelvin scale zero point is the absolute zero of physics, where all motion inside atoms ceases. Zero degrees K = −273.4 degrees centigrade.] When a *thermal electron* passes close to an ion, it is deflected from its course and can radiate away some of its energy as radio waves or other forms of electromagnetic radiation.

The nonthermal process involves cosmic-ray electrons, traveling at close to the speed of light (see Appendix, p. 244), encountering magnetic fields which everywhere permeate space between the stars. These particles spiral around in the field and radiate away some of their energy in the form of radio waves. Depending on the energy of the particle and the strength of the magnetic field involved, this process can produce radio waves, light waves, or X-rays. Another name given to the process is *synchrotron emission,* in view of the fact that such radio waves are also produced by the particle accelerators known as synchrotrons used by physicists to explore the basic constituents of matter.

## Radio Spectra—Identifying the Emission Mechanism

In order to tell whether the radio source is thermal or nonthermal, the radio astronomer measures the intensity of the received radio waves at many widely separated wavelengths (see Appendix, p. 251, for a discussion of radio source intensities). The way the intensity of the radio emission depends on wavelength is called the *spectrum* of the radio source. The spectrum of a synchrotron source shows that it becomes brighter at longer wavelengths (as predicted from theoretical understanding of the process and established from observations of

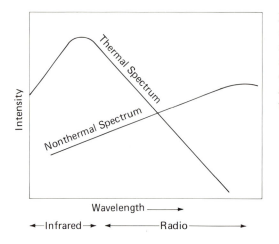

**FIGURE 1.4.** Comparison of the shapes of the spectrum for a thermal and a nonthermal radio source. The peak in the thermal (blackbody) spectrum occurs at a wavelength which is related to the temperature of the emitter.

sources in which magnetic fields are known to exist). The shape of this spectrum is shown in Figure 1.4. The brightness of a thermal source, on the other hand, decreases with increasing wavelength. Determination of the brightness of a radio source at several radio wavelengths allows its spectrum to be determined, and this is usually enough to show whether the radio source is thermal or nonthermal (synchrotron) in nature. This, in turn, allows the physical conditions in the source, such as temperature, density, and magnetic field strength, to be determined.

Nonthermal sources include quasars (Chapter 2), radio galaxies (Chapter 3), and exploding stars (Chapter 8). Thermal sources include the sun (Chapter 12) and clouds of hot gas (Chapter 8) that surround young stars.

## Position Measurement and Angular Accuracy

In order to identify the source of the radio waves, radio astronomers need to determine an accurate radio source position so that they can compare this with the optical photograph of the same part of the sky. This can sometimes be relatively easy and other times extremely difficult, depending on how faint or bright the source is and how large it appears against the sky. A faint, small-angular-diameter source, such as a very distant galaxy, may be very difficult to pinpoint, while a source that is both large and bright, such as a relatively nearby nebula in the Milky Way, is usually easier to associate with its optical counterpart.

Positional accuracies are usually measured in arcseconds. To appreciate the smallness of an arcsecond, imagine looking at the face of a dime two miles away. The fraction of your panorama filled by the dime is one arcsecond. The largest radio telescope system, the Very Large Array (known as the VLA and discussed in Chapter 17), can discern detail down to about a tenth of an arcsecond.

(Our dime would be about 20 miles away for it to cover an angle of one-tenth of an arcsecond.) In the early 1960s this was utterly beyond the technological capability of radio astronomical instruments.

The ability to discern detailed structure depends on the beamwidth, or the *resolution,* of the radio telescope. The resolution or beamwidth is given by the wavelength of the radio signal being observed divided by the diameter of the reflector. Thus the 91-meter (300-foot) diameter radio telescope of the National Radio Astronomy Observatory in Green Bank, West Virginia, operating at 20-cm wavelength, has a 10-arcminute beam, which means it can "see" structures down to 10 arcminutes across. Anything smaller than this will be blurred by the resolution of the dish and will appear to be 10 arcminutes in diameter. For comparison, the human eye cannot distinguish anything smaller than 20 arcseconds across. This limit is determined by the wavelength of light (about $5 \times 10^{-5}$ cm) divided by the diameter of the pupil (about half a centimeter). In practice, the eye's lens is not perfect and sets a limit on our capacity to see details to about one arcminute.

## Seeking New Knowledge

Radio astronomy, like all science which seeks answers beyond the borders of the unknown, requires a great deal of thought and effort and, especially recently, significant amounts of money. In asking governments for funds to construct a new radio telescope, the modern explorers of space are following a time-honored tradition. Voyages of discovery have always been costly affairs, usually sponsored by monarchs, business interests, or empires. Even Columbus needed a "research grant" from Queen Isabella to carry him across the ocean. Today our tax money is used to build the vessels of discovery, but the scientist/explorer's challenge is more subtle than it once was, and the "return on investment" is not so clear. For instance, the primary source of funding for radio astronomy in the United States is the National Science Foundation, acting on behalf of the government.

In ancient times the sponsor had an expectation that the ship would return with a cargo of spice, gold, or silver—something that could be used in barter. It is no longer so. The new explorer searches for knowledge—subtle, etheric knowledge. Such knowledge may be returned in the form of a radio image of a distant galaxy (Figures 1.2 and 1.3) or of the invisible center of our galaxy (Figure 1.5). Of course, it is impossible to attach financial worth to such images, just as it is impossible to attach value to that elusive substance called knowledge. These pictures of radio sources are beautiful in their own right, revealing the existence of previously unknown phenomena, knowledge of which broadens our perspectives about the universe into which we are born. When such scientific knowledge is woven into a fabric to produce a comprehensive view of our world, it enriches our lives.

The human race looks out into space and discovers marvellous beauty, a

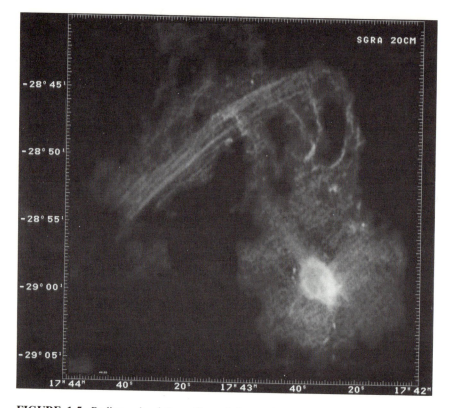

**FIGURE 1.5.** Radiograph of the radio-emitting region associated with the center of the Milky Way. A system of filamentary structures typically 20 arcseconds across, or about 3 light-years at the distance of the galactic center, crosses the mid-plane of the Milky Way at right angles. The disk of the Milky Way crosses this image approximately diagonally from upper left to lower right. A "halo" of radio emission, some 8 arcminutes or 75 light-years across, surrounds the bright radio source at the lower right, known as Sagittarius A and associated with the very center of the Milky Way. (NRAO. Observers— F. Yusef-Zadeh, M. R. Morris, and D. R. Chance.)

beauty which often lies beyond our normal powers of perception; yet a beauty which touches us as profoundly as any terrestrial sunset, symphony, or songbird. In radio astronomy the beauty is perceived by fully harnessing our imagination as we travel beyond the senses. In the following pages, you will join in the adventure and share the excitement of exploration as we journey into the invisible universe, a universe revealed by tiny amounts of radio energy reaching us from space. It has been estimated that all the radio waves ever detected by all our radio telescopes only contain as much energy as a falling snowflake. This illustrates why we need large radio telescopes and demonstrates the subtlety of the human ability to find knowledge in the elusive.

# Part II

## Extragalactic Radio Sources

# 2

# Quasars

## Journey to a Quasar

Our journey into the invisible universe starts at the quasars, very distant sources of radio waves whose nature was a complete mystery when they were first discovered in the mid-1960s. In order to see quasars up close, we have to travel a very long way out into space, and very far back in time, to an age when the universe was young and galaxies were mere fledglings on the cosmic scene. The journey must take place in our imagination, and will be assisted by what radio telescopes, connected to sensitive radio receivers and the world's fastest computers, have revealed about these cosmic sources of radio waves.

So let us travel out beyond the earth's atmosphere, past the moon, and hurtle on to leave Mars and Jupiter, mere planets orbiting our private star, far behind. Travel past Neptune and beyond the boundary of the solar system, through the cloud of distant cometary material believed to exist somewhere between the sun and the nearest stars. Continue on, past our stellar neighbors several light-years from us. Fly through the collection of stars that humans call the Milky Way galaxy, a disk 1000 light-years thick, out into the voids of space between the galaxies.

Our journey continues beyond a few million light-years (the extent of our Local Group of galaxies) out into even emptier voids of space between clusters of galaxies, then past those distant clusters which exist as much as hundreds, even thousands, of millions of light-years from earth.

Now we see galaxies whose light has been traveling toward earth ever since primitive life crept over our planet, light which has hurtled relentlessly through space at 300,000 km/s, day after day, year after year, for hundreds of millions of years, to reach us.

Our journey to the quasar must go even further, past most of the clusters of galaxies visible to the largest optical telescopes on earth. We must travel beyond the distances over which earthbound optical astronomers can see. It is out there, billions of light-years away, that radio telescopes "see" most of the quasars, intense beacons of radio energy which signal remarkably violent forces at work. At those distances we are looking a billion or even 10 billion ($10^{10}$) years back in time.

As we move closer to the quasar, we see a bright core of radio and light emission. The core radiates the energy of a thousand billion ($10^{12}$) stars, yet the source is relatively but a pinpoint in space (Figure 2.1). How can this extraordinary luminosity, so much energy, be created in such a small volume of space? That is one of the fundamental and continuing mysteries of quasars.

Astronomers are trying to explain some of the mystery. They have struggled with the physics yet nothing quite fits. But at least they have some inkling about what might be going on. The lack of theoretical understanding does not prevent us from exploring further, and so we travel closer.

In the newly born universe, before stars cooked up the elemental constituents of planets and living things, there were swarms of newly born, first-generation stars which gathered in vast clouds. These were the first galaxies. Within those

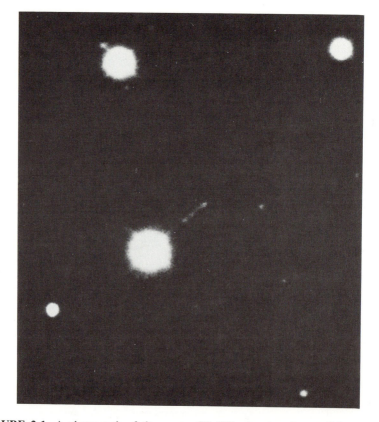

**FIGURE 2.1.** A photograph of the quasar 3C 273, showing the small jet emerging from a bright nucleus which otherwise looks very much like the foreground stars seen in the photograph. (National Optical Astronomy Observatories.)

galaxies the largest stars grew rapidly and in vast gulps sucked energy from their massive hearts. The stars contracted and burst forth in death throes that shed vast clouds of atoms to seed space with the stuff of life, atoms later used in the creation of living things. The remnants of the stellar cores sometimes collapsed inward at the speed of light, to be extinguished in catastrophic minutes of time.

The remains of some such stellar collapse may be a *black hole,* formed by forcing enormous amounts of matter into a small volume, so small that the force of gravity near the object becomes so intense that nothing can escape its pull. A black hole is not really a hole in space. It is a highly concentrated blob of matter which is invisible because no light can escape from it. Any light that comes too close will be sucked in, to travel no further.

It is theorized that in that early universe black holes drifted where stars had once been. Other stars inexorably joined in the death rituals, and matter swelled and blew around in cosmic fury, black hole pounding black hole, absorbing each other, growing in mass until a hundred million stars' worth of material was gathered together at the heart of the galaxy. Still the fiery furnaces roared and matter plummeted inward. Some material was blasted out, energized to the speed of light, producing a flare of energy shining brightly across all of space and time, signaling terrestrial astronomers that something extraordinary occurred a very long time ago, near the edge of our universe.

Such is probably the heart of the quasar, the intensely luminous core of very distant galaxies. This is what the radio telescopes on earth now see as they peer into one of nature's primal mysteries, witnessing phenomena beyond the wildest dreamings of a Galileo, a Kepler, or even mid-twentieth century astronomers. Those signals have revealed most of what we know about our universe in its earliest times.

We are now close to the nucleus of a galaxy which existed five billion years ago. In front of us a massive black hole is spinning as steadily as a gyroscope. Around it is a vast disk of interstellar matter where the remains of dead stars swirl violently. When the gravitational pull of the black hole overcomes the resistance provided by centrifugal motions, some of the hot matter falls into the clutches of the massive, invisible object. Violent collisions cause the gas to be heated to high temperatures and then it mostly falls into the black hole. But near the poles of the spinning hole some gas suddenly finds that instead of plummeting into the void of cosmic nothingness, it is blasted outward in two directions, streaming along the black hole's axis of rotation with such force that the gas is accelerated close to the speed of light.

These gases, heated to great temperatures, are propelled straight outward in long jets which are so energetic that nothing sways them from their initial path. They stream out into space, long, luminous structures which shine brightly in the radio band (Figure 2.2).

For a hundred thousand years, a million years, sometimes for a hundred million years, these jets stream out into space, pushing matter out ahead of them, inflating gigantic lobes of radio emission beyond their ends like cosmic

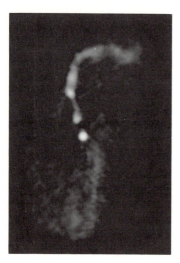

**FIGURE 2.2.** Radiograph of the quasar 2300–189, located about $9 \times 10^8$ light-years away, observed with the VLA at a wavelength of 21 cm. The very luminous core appears as a fuzzy patch of light at the center while two curved, precessing jets, each about half a million light-years long, reach into space around the quasar. Vertical size is 2″. (NRAO. Observer—J. J. Condon.)

balloons. These balloons expand and within them the electrons traveling near the speed of light spiral around the magnetic fields and emit radio waves by the synchrotron process (discussed in Chapter 1). Five billion years later those radio waves reach a radio telescope on earth where they are absorbed to cause a minute current in an electronic circuit to indicate that they have arrived. The current is amplified and recorded. The five-billion-year-old radio energy has been converted into a number, a squiggle on a piece of paper!

The earth is continually bathed in a radio glow from all of space, including the signals from thousands of quasars. Those radio waves strike the ground and are absorbed. Their combined energy is minuscule compared to the light from the sun.

## Quasi-Stellar Radio Sources

Quasars are a marvellous example of the extraordinary cosmic wonders revealed by modern science. They include some of the most luminous and most distant objects in the universe. Using remarkable radio telescopes which span continents (see Chapter 17), radio astronomers have, for the first time, looked into the violent, dynamic, living cores of the quasars. From the distances of millions of light-years that separate us from those caldrons of raw primal energy, we penetrate into the quasar and witness a dance of creation and destruction that determines the fate of galaxies.

To telescopes on distant earth, quasars appear as pinpoints of radio and light emission. Superficially they look just like stars (Figure 2.1). In the early 1960s, when quasars were first identified, their optical images were confused with those of ordinary stars. Originally the peculiar radio emitters were labeled *quasi*-stell*ar* (starlike) radio sources, a name which was quickly contracted to *quasar*. Some of the quasi-stellar objects are radio quiet and they are sometimes included under the generic label of *QSO*.

## Parent Galaxies

Controversy simmers as to whether quasars are isolated objects that are not associated with clusters of galaxies, or whether they are the cores of otherwise "normal" galaxies undergoing violent explosions. In the latter case, surrounding galaxies may have been missed because the quasar so dramatically outshines them—a quasar may shine 100 times more brightly than a galaxy containing $10^{11}$ stars. (If a quasar phenomenon occurred at the center of our Milky Way, located 30,000 light-years away, the quasar would appear as bright as the moon.) Painstaking research by many astronomers has revealed that some quasars are indeed enormously luminous explosive nuclei of elliptical galaxies. (An elliptical galaxy appears to contain only closely packed stars and shows no spiral patterns such as are visible in the galaxy shown in Figure 1.2.) These explosions are so bright that because of the glare, we can barely see the surrounding galaxy. Only through use of very sophisticated photographic techniques and the largest optical telescopes have the parent galaxies been revealed in some of the closer quasars. Furthermore, because of the enormous distance to such objects, the surrounding cluster of galaxies remains, as yet, invisible.

## Identification of Quasars

In 1918 a galaxy known as Messier 87 (which is 33 million light-years away in the direction of the constellation Virgo) was photographed and revealed a surprising jet of luminous matter shooting out of its interior (Figure 2.3). In the late 1950s, as radio astronomy moved into adolescence, and well before the giant radio telescopes of today had been constructed, radio signals from M87 were discovered. It was one of the first examples of a mystery never before encountered—the radio galaxy. Radio emission from these radio sources has been observed to originate as far as 10 million light-years from the optically visible galaxy. An early radio map of M87, superimposed on the optical image, is shown in Figure 2.3. (See Chapter 3 for a description of such radio maps.)

In the 1950s other peculiar objects were identified with some of the strongest radio sources in the sky. The Crab nebula, the remains of a star that exploded back in 1054 A.D. (see Chapter 8), was found to be a powerful source of radio waves. A peculiar galaxy in the constellation Cygnus was later associated with the radio source called Cygnus A (Figure 1.2.).

As radio astronomy technology improved, especially when groups of radio dishes were connected together to act as single large radio telescopes (to be described in Chapter 17), many weaker radio sources were discovered and their positions measured. Yet the first positions were still too inaccurate to

20 2. Quasars

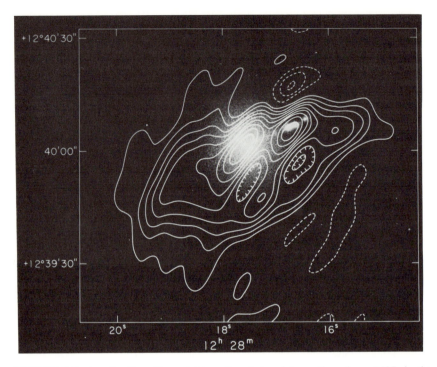

**FIGURE 2.3.** A radio ''map'' overlaid on the photo of the radio galaxy, M87, in the constellation of Virgo, one of the first radio sources to be identified with a distant galaxy. The optical image of the radio-emitting jet is clearly visible in the photograph to the right of the central galaxy. (NRAO. Observers—D. E. Hogg, G. H. McDonald, R. G. Conway, and C. M. Wade.)

allow optical astronomers to identify the associated object, except in rare cases where something very peculiar was visible at that location. Only then did astronomers feel confident in relating the source of radio waves to the optically visible object.

In 1963 the moon moved in front of the radio source known as 3C 273. Because astronomers knew the moon's position very accurately, they simply timed the disappearance and reappearance of the radio source as it was blocked (*occulted*) by the moon and converted the timing information into a position on the sky. The observations had to be repeated during another lunar occultation in order to resolve position ambiguity, because two points on the spherical silhouette of the moon could produce the same time of disappearance. The second occultation would occur when the moon traveled along a slightly different path and hence the timing information of two occultations could be combined to give an unambiguous position for the radio source. In this way the location of radio source 3C 273 was accurately pinpointed and the 200-inch telescope on Palomar mountain was able to photograph the object precisely.

The photos produced an incredible discovery: 3C 273 appeared to be a star. However, much to the surprise of all involved, this one had a luminous jet protruding from it (Figure 2.1). Radio signals seemed to originate from the center of the star as well as the tip of the jet. This starlike object looked like M87, but did not appear to be a galaxy. This was one of the first quasars unambiguously identified. Soon a starlike object was also found at the position of the radio source 3C 48.

As a consequence of these identifications, it was briefly believed that many radio sources might be very peculiar and mysterious "radio stars" in our own galaxy. (The sun, a normal star, is also a source of radio waves, but at a great distance outside our solar system it would hardly be noticeable.) When astronomers studied the light from the suspected "radio stars" they found that they were not stars at all. Analysis of the light showed that these objects had large redshifts (see Appendix, p. 250), which implied that they were located very far away, as far as, and often even farther than, most visible galaxies. By 1963 the mystery of the quasars had crashed about the astronomical community in full force. What sort of object could look like a star (even if it had a jet protruding from it), yet be farther away than most known galaxies? What could possibly be emitting so much energy as to shine more brightly than any other object in the universe?

## Quasar Redshifts

When optical spectral lines are observed in distant galaxies (or quasars) they are found to occur at a wavelength different from that measured on the same atom studied in the laboratory. This is due to the Doppler effect (discussed in the Appendix) and indicates that the observed galaxies are all moving with respect to us. In fact, all but some of the closest galaxies appear to partake in this universal recession, or redshift, and this phenomenon gave rise to the idea that we live in an expanding universe. The redshift of a galaxy is therefore an indication of its distance.

The optical spectra of quasars show large redshifts, which places them very far away—unless it is argued that the redshifts are due to some other phenomenon. For example, it is possible that the strength of gravity in the quasar core is so large that it produces what is known as a gravitational redshift, and for some years that was a major point of argument. However, since the discovery of the quasar parent galaxies, the gravitational redshift hypothesis has lost most of its steam. In addition, the faint light from many parent galaxies has now been independently measured and found to be redshifted by the same amount as the light from the associated quasar, so most astronomers believe that quasars are indeed located at very great distances from us. However, this is still not certain, as the discussion on galaxy–quasar pairs (below) illustrates.

Some of the most distant quasars are receding from us at 90% of the speed of light, placing them 10–15 billion light-years distant, therefore up to 15 billion years away in time. (The uncertainty is due to a related uncertainty in the relationship between redshift and distance—known as the Hubble constant, assumed to be 30 km/s per million light years in this book—determined from observation of nearby galaxies.) Thus they existed at an epoch when the universe was perhaps only a few billion years old, because the current age of the universe is believed to be somewhere between $15 \times 10^9$ and $20 \times 10^9$ years. Quasars typically recede at 50–90% of the speed of light.

There are at least 2000 known quasars. The most luminous quasar—that is, the one emitting the most energy, known as S500 14+81—is also one of the brightest in the sky and easily visible to a medium-sized optical telescope, even though it is $10^{10}$ light-years away. It outshines the entire Milky Way by 60,000 times, yet all its energy is emitted from a tiny core less than a few light-years across.

## Brightness Variations

In the early days (1960s) of quasar identification, the challenging problem was not just discovering the distance to the quasar, but ascertaining over what volume of space the intense radio emission was being generated. A startling answer was produced in a most unexpected manner when astronomers discovered that the light from quasar 3C 273 changed in brightness over a year. Furthermore, a subsequent search of old photographs of the region of sky around 3C 273 showed that it had undergone sudden changes in brightness over a period of 80 years. This incredible discovery was completely unexpected. First, astronomers seldom observed time variability in objects other than stars within our galaxy. Secondly, quasars are intensely luminous and are a problem in physics even if the energy is generated over volumes of space and galactic dimensions. This problem becomes far worse if the emitting region is very small. Variability of quasar brightness showed that the source had to be of the scale of a few light-years across. The reason is as follows.

The speed of light is a basic speed limit in our universe. Knowing that an astronomical object's brightness varies on a time scale of a year immediately gives its size as less than or equal to one light-year across. If the object were, say, ten light-years across, there is no way that the impulse causing it to brighten could travel all the way across the source in less than ten years. Therefore, significant time variability of the brightness of such an object would tend to occur over this time period.

Soon after the discovery of optical variability in quasars, radio astronomers began to monitor them for radio variability, something which seemed absurd at first, because most of the total energy from quasars seemed to be coming from a radio source, and it, surely, could not be as small as the size implied by the optical brightness variations. It was found that radio emission from quasars varied from year to year, meaning that the luminous radio-emitting regions were not merely very bright, but also very small. This posed a tremendous problem for the theorists: how to explain huge quantities of energy being emitted in bursts from a tiny volume of space. Radio sources are now known to vary on most time scales from years, down to months, and even from day to day.

## The Most Luminous Objects in the Universe

The full mystery of the quasar becomes more profound when all the information is considered together. They are very far away and the luminous cores are very small. At the distance inferred from its redshift, a quasar is emitting the energy equivalent to $10^{13}$ stars like the sun, all the energy being generated in a volume of space not much larger than the solar system. The luminosity of quasars would not be so much of a problem if they were located relatively close to the sun, somewhere in the Milky Way, for example. (You can see the light from a small flashlight from across a football field but you can't see it 10 miles away because it simply is not luminous enough.) For a closer distance the quasar luminosity would be smaller by an amount which depends on the distance squared. (Bring the quasar twice as close and it will need only one-fourth the luminosity to appear as bright as before.) If quasars are actually closer, their energy generation would not be a problem, but then their large redshifts would suddenly be a pressing mystery and their association with galaxies inexplicable.

The quasar's intrinsic luminosity, the amount of energy it actually radiates (estimates of which depend on the distance assumed for the quasar), is the most important property astronomers have to explain, because an enormous amount of evidence supports the interpretation of redshifts as being a reliable distance indicator.

## Galaxy–Quasar Pairs

In recent years, after the parent galaxies of several quasars had been recognized, it seemed as if the distance controversy was neatly resolved. Alas, this does not appear to be as neat as astronomers would like! There is still some doubt concerning the large, cosmological distance to quasars because of the apparent discovery of galaxy–quasar pairs having two different redshifts! In these cases a quasar appears to be positioned very close to, or almost coincident with, what is clearly a nearby galaxy. Unfortunately, the highly discordant redshifts of the galaxy–quasar pairs imply that they are at very different distances. For example, the galaxy known as NGC 1073 has a redshift indicating a distance of about $50 \times 10^6$ light-years, but is paired with a quasar allegedly $3 \times 10^9$ light-years away, if its redshift is a reliable distance indicator. Astronomers have tried hard to deal with these coincidences on the basis of chance alone, but the arguments have not yet been convincing enough to resolve the dilemma.

Figure 2.4 is a photograph of a galaxy, known as NGC 4319, with a quasar (Markarian 205) located immediately adjacent to it. According to their redshifts,

**FIGURE 2.4.** The galaxy NGC 4319 and quasar Markarian 205, the bright object just to the south of the larger galaxy. The halo around the quasar and light from the galaxy appear to blend, suggesting that these two objects are directly associated with each other. However, according to the redshift the quasar is $1.2 \times 10^8$ light-years away, 12 times farther away than the galaxy. (J. W. Sulentic.)

the quasar is located 12 times further away ($120 \times 10^6$ versus $10 \times 10^6$ light-years) than the galaxy. A bridge between the galaxy and the quasar appears on photographic images, and in spite of the fact that the light levels are so low, the discovery seems to have been verified. This pair of objects is a definite thorn in the side of the usual interpretation of quasar distances. However, in view of the very recent discovery of gravitational lenses (see Chapter 14), it may not be so unusual to find quasars in coincidental alignment with relatively nearby galaxies. We will have to wait for further observations before this apparent mystery is resolved.

Astronomers have also found close pairs of quasars with very different red-shifts. Furthermore, it is known that several groups of very closely interacting galaxies contain members whose redshifts are completely at odds with the other members. This further confuses the simple view of quasar, or even galaxy, redshift interpretations.

The arguments about redshift interpretation and the distance to quasars continue, although the balance of evidence is now heavily in favor of large distances to quasars. However, other fantastic objects, the radio galaxies, are clearly relatively nearby. They, too, release enormous amounts of energy over time spans of millions of years.

# 3

# Radio Galaxies

## The Largest "Things" in the Universe

Quasars have tantalized us. Because of their extreme distance, it is difficult to see deep enough into their hearts to learn all their secrets. Very likely closer to home, however, the radio galaxies, which include some of the largest objects in all the universe, exist in large numbers. Somewhere inside these galaxies is a source of energy which feeds a double radio source which may be as large as 15 million light-years across: for example, in 3C 236, the most gigantic radio source of all. Radio galaxies are now believed to be a more recent manifestation of the same phenomenon that gave rise to quasars.

Some of the earliest (1960s) observations of distant radio sources revealed that the radio emission was usually coming from two regions in space located on opposite sides of the optically visible galaxy. The double radio source might be a minute of arc or so in extent with an optically visible galaxy located between the radio "blobs." This observation was supported as more smaller and weaker radio sources were studied. Double radio sources were found to be extremely common, but due to the poor resolution of early radio telescopes little more could be said than that the radio source was a double. It became fashionable to explain these radio sources as explosive events inside galaxies; which for some reason ejected material in two directions. The central galaxies were often observed to have very active nuclei, active in the sense that very violent motions were occurring there, as inferred from the nature of their light emission—their light spectra showed Doppler shifts which implied very chaotic motions.

Recent radio observations of the double radio structure showed that processes inside these galaxies are far more complex and fascinating than simple explosions of entire galaxies! In the radio galaxies immense amounts of radio, light, and X-ray energy are generated by dramatic events in the nucleus of what is usually the most massive member of a dense cluster of galaxies. Matter is indeed streaming outward from the galaxy to create amazing displays of cosmic fireworks. Figure 3.1 shows the radiograph[1] of a typical radio galaxy, this one known as Hercules A, or 3C 348. Two *jets* lead toward filamentary lobes of radio-emitting material, while a distinct ring of emission, which suggests a complex interaction with surrounding intergalactic gas often found in clusters of galaxies, can also be seen. The intergalactic material is in the form of very hot, ionized hydrogen which can be observed by means of the X-rays it emits.

Radio galaxies may produce a million times more energy across the entire electromagnetic spectrum than a normal spiral galaxy, and their radio emission alone can outshine spiral galaxies by 100,000 times. Another example of a radio galaxy is shown in Figure 3.2. This object, 3C 310, appears to be blowing bubbles!

From the depths of many radio galaxies, highly elongated and stable jets appear to continually drive matter into two enormous radio-emitting regions known as the radio source *lobes*. The new view of the physics and evolution of these immense radio sources suggests that these remarkable behemoths are not actually exploding, but are continually spewing out incandescent matter in a steady stream that flows for millions, even hundreds of millions, of years, and it is suspected that this hot matter is propelled outward from a black hole at the very heart of the active galaxy.

## Centaurus A

Centaurus A is the prototypical radio galaxy. It is extremely well studied, because it happens to be the radio galaxy closest to us, $15 \times 10^6$ light-years away. It is a classic example of its species, and because of its proximity, appears huge on the sky. Images of the radio source are shown in Figure 3.3 as a series of contour maps. Just as geographical maps use contour lines to join points with the same elevation, so a radio contour map joins points where the intensity of radio emission is observed to be the same. A series of closed contours indicates a peak in the radio emission, which, in radiographs such as Figure 3.2, would appear as a bright part of the image. Radio contour maps are extensively used by radio astronomers to analyze their observations because the astronomer can accurately read relevant data from the contours. These maps are, of course, drawn after extensive computer processing.

The radio source Centaurus A is associated with an elliptical galaxy, known as NGC 5128 (shown in Figure 3.4), which is 7 minutes of arc, or 30,000 light-years, across. (An elliptical galaxy differs from a spiral galaxy in that it

---

[1] Many of the radiographs, such as Figure 3.1, were made with the National Radio Astronomy Observatory's Very Large Array radio telescope, located near Socorro, New Mexico, which uses 27 75-foot-diameter radio dishes to simulate a single radio telescope 21 miles in diameter (see Chapter 17). It is capable of a resolution of around one second of arc (depending on the wavelength), as good as that of the world's largest optical telescopes. Similar radiographs produced with antenna systems in Westerbork in the Netherlands and at Cambridge, England, also appear in this book.

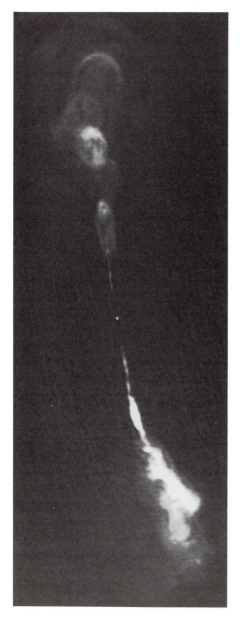

**FIGURE 3.1.** The 6-cm wavelength radiograph of Hercules A (3C 348). This is one of the brightest and most luminous radio galaxies in the sky. Two jets streaking away from the tiny faint nucleus reach half a million light-years into space. The eastern jet (bottom) shows ''wiggles'' indicative of unstable flow of radio-luminous matter while the western jet (top) shows several distinct ''rings.'' Resolution of half an arcsecond. Horizontal size is 1.2′. (NRAO. Observers—J. W. Dreher and E. D. Feigelson.)

consists of densely packed stars with very little matter between them.) As can be seen in Figure 3.4, however, NGC 5128 is, in the mid-nineteenth century words of Sir John Herschel, ''cut asunder . . . by a broad obscure band.'' There is indeed an obscuring band of interstellar dust seldom, if ever, found in elliptical galaxies, but typical of spiral galaxies. The dust band even shows

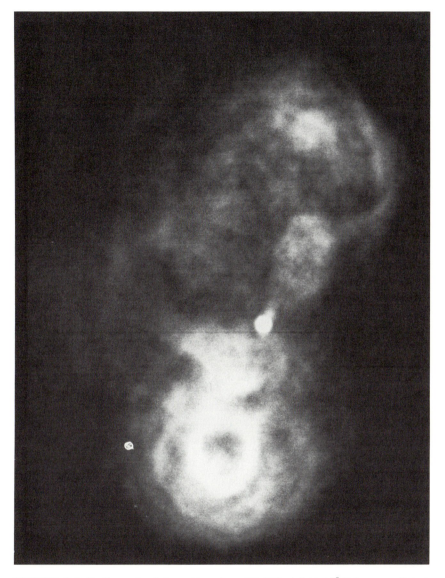

**FIGURE 3.2.** Radiograph of the radio galaxy 3C 310, $4.5 \times 10^8$ light-years distant, which appears to be blowing "bubbles." This image was produced at 21-cm wavelength with a resolution of 4 arcseconds. The bright core is associated with an elliptical galaxy whose size is approximately that of the core spot. The extent of the radio-emitting region is $6 \times 10^5$ light-years across. Vertical size is 5.5′. (NRAO. Observers—W. van Breugel and E. B. Fomalont.)

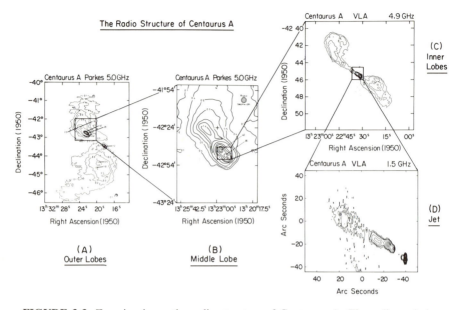

**FIGURE 3.3.** Zooming in on the radio structure of Centaurus A. The radio emission reveals a series of lobes of more intense radiation. The overall extent of the largest lobes (A) is $1.5 \times 10^6$ light-years while the small jet at the core of the galaxy (D) is about $10^4$ light-years long. Figure 3.4 shows the radio source on the scale of the upper right frame. (NRAO. Observers—J. Burns, E. D. Feigelson, and E. J. Schreier.)

rotational motions characteristic of spiral galaxies. Since the 1950s it has been suspected that NGC 5128 might represent the merger of two galaxies, and now astronomers believe that Centaurus A may, indeed, be a very public example of galactic cannibalism—an elliptical galaxy gobbling up a spiral! In the process the galaxies suffer indigestion and clouds of radio-emitting material are impolitely belched out.

## The Structure of Centaurus A

In Figure 3.3 the radio emission from the lobes can be seen to extend over 7 degrees of sky, equivalent to $1.5 \times 10^6$ light-years. The sequence of contour maps in this figure shows a series of higher-resolution views, or a "zoom," on the radio source. The inner region contains a smaller double radio source (the inner lobes) whose contours are superimposed on the optical photograph in Figure 3.4. The two small lobes point in the same direction as the larger-scale structure. At the very core of NGC 5128 a jetlike feature is evident. It appears that a series of ejections of material must have occurred along the same directions in order to produce the structures seen in Figure 3.3.

Detailed observations by many astronomers have now revealed that the large, bright, elongated region to the upper left in Figure 3.4 extends to 50,000–

**FIGURE 3.4.** The 6-cm radio contour map of Centaurus A superimposed on an optical photograph of the galaxy associated with NGC 5128. The inner lobes of the radio source jut out at right angles from the dark dust lane which crosses this unusual galaxy. About $1.5 \times 10^7$ light-years away, Centaurus A is the closest radio galaxy in the sky. (NRAO. Observers—J. O. Burns, E. D. Feigelson, and E. J. Schreier.)

65,000 light-years out and includes three large, optically visible nebulae. Chains of hot blue stars have also been observed in this region. This means that stars have recently formed in a part of space obviously well removed from the galaxy itself. This phenomenon of star formation being triggered near active galactic nuclei has since been observed in other galaxies. In another region of space, some 130,000 light-years from NGC 5128, diffuse gas, filaments, and X-rays have been observed.

The inner radio lobes are symmetrically located around the galaxy and extend about 30,000 light-years from the core. They generate no visible light or X-ray emission. At the center is the small jet, four light-years long and broken up. It shows associated X-ray emission. Whatever is being beamed out from the core retains its straight path from 4 to 20,000 light-years out, nearly equal to the distance from the earth to the center of our galaxy (30,000 light-years). However, no such long, straight features are found in the Milky Way. At the very nucleus, or core, of NGC 5128 there is a strong radio source whose detailed structure has not yet been probed.

The radio source is associated with X-ray emission due to a gas believed to be at $10^7$ K. Less intense X-ray emission comes from an envelope that extends

10,000 light-years from the galaxy: the gas's temperature is $2 \times 10^7$ K and its mass is as much as $2 \times 10^8$ solar masses. It is this surrounding gas, at high temperatures, which plays an important role in confining and containing the jet of material feeding the radio lobes. This phenomenon is believed to be typical of other radio galaxies which, due to their greater distance from us, cannot be studied as easily as Centaurus A.

The radio, optical, and X-ray observations of Centaurus A have given tremendous insight into the complexity of radio galaxies, but what was especially dramatic about the first radio observations of Centaurus A was that they confirmed that radio waves from extragalactic radio sources are generated by the synchrotron emission process. This confirmation came from the observation that these radio waves are highly polarized.

## Polarization

Relativistic electrons, those traveling near the speed of light, will spiral around magnetic fields and in the process produce radio signals by synchrotron emission (see Chapter 1). This radiation also exhibits a property called *polarization*. That means that the radio signals vibrate in some preferred direction.

(Imagine a rope held at two ends. Flip one end up and down and a wave travels down the rope. That wave is vertically polarized. Hold the rope taut and flip it sideways. A horizontally polarized wave now travels along the rope.)

In the case of synchrotron emission it was predicted that the radiation would be polarized, that is, show a preferred direction of vibration, provided that the magnetic field, about which the electrons spiral, is at all organized. The radio wave is expected to be polarized at right angles to the magnetic field. Indeed, such polarization was discovered in the radio waves from quasars and radio galaxies, showing that the magnetic fields in these objects are often highly organized.

The radio telescope can detect the angle of polarization of the incoming radio wave, from which the field direction at the source can be inferred. But a slight complication is present in all cases. As the radio wave passes through space on its way to the radio telescope, it suffers what is called *Faraday rotation;* that is, a rotation of the angle of polarization as the radio wave traverses a medium containing a magnetic field and thermal electrons (see Chapter 1). Interstellar space in our galaxy and the space within and around distant radio sources contains magnetic fields and thermal electrons. Therefore, Faraday rotation is produced in the radio source and in our galaxy. It is possible to make a series of observations which allow astronomers to correct for this effect, which means that they can infer the angle of polarization, and hence the magnetic field direction, at the source. In the jets of many radio galaxies the magnetic field is aligned along the direction of the jet itself, and at the outer edge of the radio lobes the field is often parallel to the boundary of the lobe, implying that the field has been pushed up against this boundary.

## Cygnus A

In 1953 the galaxy associated with Cygnus A, the second brightest radio source outside our solar system (the sun is the brightest radio source in the sky and the Cassiopeia A source the brightest outside the solar system), was identified. The most recent radio map of this source is shown in Figure 3.5. This Very Large Array image shows the most astonishing details. The radio lobes manifest beautiful diaphanous filaments whose subtle patterns belie the amazing energies associated with this source. A faint yet stunning radio jet, less than a tenth of a percent as bright as the lobes, can be seen heading toward the northern lobe (which was shown blown up in Figure 1.2). The radio double is centered on a peculiar galaxy which was originally believed to be two galaxies in collision. In the 1950s two famous astronomers, Walter Baade and Rudolph Minkowski, argued about this, and bet a bottle of whisky or a thousand dollars, depending on whose version of the story you believe, on whether or not Cygnus A was a colliding galaxy. The issue was settled—against the colliding galaxy hypothesis— when it was realized that double radio sources were too common to be explained by intergalactic collisions. However, since then the explanation is again in question. Galactic cannibalism, a variation on the theme of simple collision, may be at work in many, if not all, radio galaxies.

Several "hot spots" can be seen in the radio lobes. These hot spots are characteristic of many double radio sources and are often found at the end of the line of the jets, where material crashes up against the boundary separating the radio lobe from intergalactic matter.

## Other Remarkable Species in the Cosmic Zoo

Each radio source appears unique, yet underlying patterns emerge. They all show lobes of extended emission, far removed from the central galaxy. Most of them show jets, sometimes one-sided, sometimes two-sided. These jets are sometimes bent or swept back, which seems to indicate motion through surrounding intergalactic gas in the cluster in which the radio galaxies occur.

3C 75 is the spectacular object shown in Figure 1.3. Four enormous radio jets blast out from the central galaxy which lies at the center of a dense cluster of galaxies about $3 \times 10^8$ light-years away. The object appears to have two cores, each emitting two jets which swirl and twist through space, apparently swept back by the motion of the galaxy (or interacting galaxies) through surrounding intergalactic material. The jets appear to be intertwining in the northern parts. The power associated with 3C 75 is 100 million times the energy output of our sun.

The extent of the jets billowing out of 3C 75 is enormous—a million light-years across. The awesome amount of energy associated with these jets, which have long since been expelled from the cores of their galaxies, seems to imply that energy sources other than the original explosive energy of ejection must

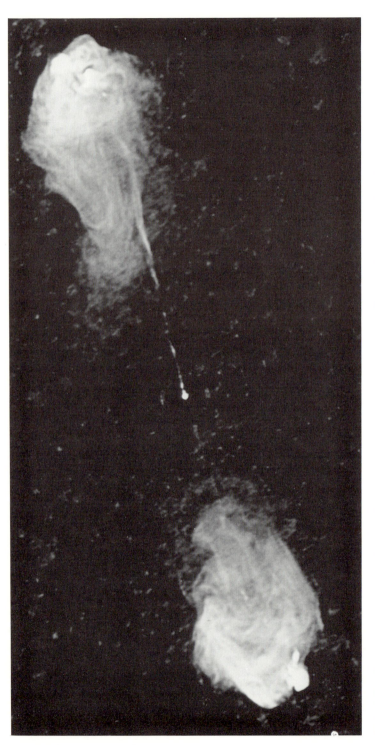

**FIGURE 3.5.** The powerful extragalactic radio source Cygnus A, one of the first to be identified with a galaxy. This 6-cm wavelength radiograph reveals fine filamentary structure in the two radio lobes, separated by about $3 \times 10^5$ light-years. An enlarged version of the right-hand lobe is shown in Figure 1.2. A giant elliptical galaxy is known to be centered at the small bright spot in the center of the picture. Thin jets connect the central "engine" in the galaxy to the powerful radio lobes. Vertical size is 1'. (NRAO. Observers—R. Perley, J. Dreher, and J. J. Cowan.)

be operating in order to keep the material emitting radio signals for so long (many millions of years). Much of the energy powering the radio source may be coming from the region of the radio lobes themselves, rather than just from the core of the galaxy, a phenomenon to be discussed in Chapter 4.

The presence of two bright cores in 3C 75, so closely spaced, may be an example of galactic cannibalism at work. In this case two galaxies may have coalesced so that their two nuclei are now very close together, perhaps about to consume each other.

NGC 6251, whose radio contours are shown in Figure 3.6, is a radio galaxy $3 \times 10^8$ light-years away with a huge radio jet, about $10^6$ light-years long, and a weaker counterjet, both of which are strikingly obvious in the contour map (made at 1664 MHz) given in the figure. The series of radio maps show the remarkable continuity from the smallest scale to the largest in this radio source. (Radiographs of this source are shown in the next chapter.)

Figure 3.7 is a contour map of 3C 449, a radio source associated with an elliptical galaxy $10^8$ light-years distant. This source is remarkable for its mirror symmetry (imagine placing a mirror vertically between the two sides of the source). Swirls in the two extended lobes mimic each other, yet they are several hundred thousand light-years in extent. For example, both jets show bends 300,000 light-years from the galaxy. Symmetry appears to be common in radio galaxies, which often show either mirror image or inversion symmetry. An example of the latter is M84, Figure 5.2. To picture inversion symmetry, imagine rotating the radio structures about an axis emerging vertically out of the plane of the figure, at the core of the source.

NGC 1265, Figure 3.8, is one of the best examples of a mirror-symmetrical radio source in which the jets have been swept back through space into great arcs, probably due to the galaxy's motion through the gas that fills the cluster in which it finds itself. This cluster gas is revealed by the X-rays it emits.

## Properties of Radio Galaxies—A Summary So Far

Powerful radio galaxies, all of which lie in galaxy clusters, are relatively close to us compared to quasars. The galaxy redshifts indicate typical distances of less than a billion light-years, with most being closer than half this distance. Quasars, on the other hand, are usually farther than this. Most radio galaxies show symmetrical double-lobed structures which range in size from a few hundred thousand light-years to $5 \times 10^6$ light-years across. Small cores, associated with the optical galaxy, are found in most radio sources. When the lobes are asymmetrical, the brighter radio-emitting lobe is usually found closer to the core. From observations of their radio spectra and polarization it is known that the extended radio-emitting regions contain magnetic fields and that their radio energy is generated nonthermally.

Symmetry in double-lobed radio sources is very striking. Whatever the reason for the lobed structure, whatever the central engine that drives the whole thing,

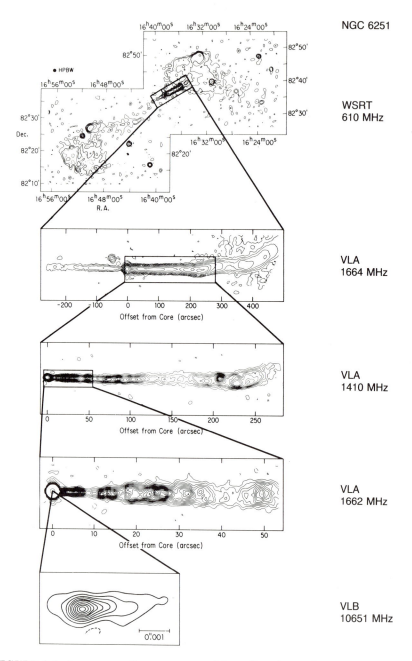

**FIGURE 3.6.** A sequence of contour maps of the radio galaxy NGC 6251. Each map, obtained at the frequency indicated, shows more detailed structure than the preceding map in the sequence. The extent of the radio-emitting region is enormous, approximately $6 \times 10^6$ light-years across. A broadening of the very straight jet as it reaches into space can be seen, and numerous knots of emission are highlighted by sets of closed contours. Compare this map with Figure 4.1, the radiograph of the same jet. WSRT = Westerbork Synthesis Radio Telescope; VLA = Very Large Array; VLB = Very Long Baseline Interferometer. (NRAO. Observers—R. A. Perley, A. H. Bridle, and A. G. Willis. Reproduced, with permission, from *Annual Review of Astronomy and Astrophysics,* Vol. 22. Copyright © 1984 by Annual Reviews Inc.)

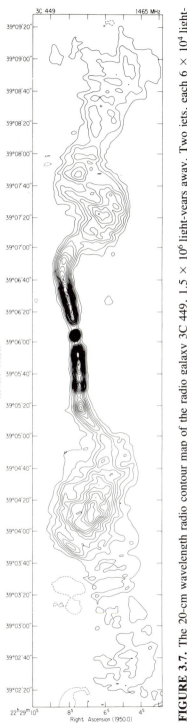

**FIGURE 3.7.** The 20-cm wavelength radio contour map of the radio galaxy 3C 449, $1.5 \times 10^6$ light-years away. Two jets, each $6 \times 10^4$ light-years long, power the symmetrical extended lobes of radio emission. (NRAO. Observers—T. Cornwell and R. A. Perley.)

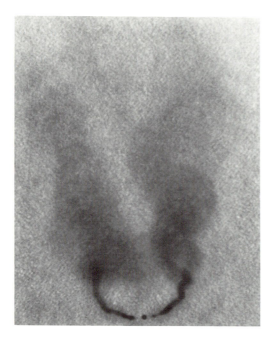

**FIGURE 3.8.** Radiograph of the galaxy NGC 1265 located in the Perseus cluster of galaxies, at a distance of approximately $2 \times 10^8$ light-years. This radiograph is presented as a negative print. The jets feeding the radio lobes have been swept back, probably as the result of the central galaxy moving through intergalactic material in the cluster. The great swirling lobes extend $1.5 \times 10^5$ light-years from the galaxy. Vertical size is 3.5'. (NRAO. Observers—C. O'Dea and F. N. Owen.)

symmetry is at its heart. What happens on one side also happens on the other. Two opposite directions are favored above all else, and in nature such preference is most likely related to something spinning about an axis, something that has a north and a south pole.

The structure of the sources appears to be correlated with the power of their emission and with galaxy type. Very large radio sources originate in elliptical galaxies rather than spiral galaxies, and large galaxies are stronger radio sources than smaller ones. The sources are usually located close to the centers of the clusters in which they find themselves. The most powerful radio emitters are typically from a quarter of a million to one million light-years across, and the edges of the radio-emitting region are brighter than the inner parts (edge-brightened). Hot spots are regions of greatest intensity of radio emission and lie along the outer parts of the source. The powerful radio emitters rarely show jets, perhaps because the jets are so much fainter than the radio lobes.

Weaker sources are generally complex and far larger, up to $15 \times 10^6$ light-years across. They are edge-darkened, which means that the brightest emission comes from the central parts of the lobe, and may often contain two-sided jets which may be more likely to show curvature on the sky. The weaker sources consist of the usual double radio lobes, but their central cores are often very bright, making them look like triple sources. Intermediate-power radio galaxies may exhibit a one-sided structure or have wide-angle tails, as opposed to straight sources or swept-back trails.

Jets are long, straight structures which emerge from the core and point into the lobes, and appear to channel energy to power the lobes. In the next chapter we will next take a closer look at these jets and explore the central engine that powers the cosmic light and radio show that is a radio galaxy.

# 4

# Cosmic Jets, Black Holes, and Cannibalism

## Cosmic Jets

Jets have been found in over 125 radio galaxies and quasars. Long, narrow streams of highly energetic gas squirt from the center of a galaxy, emitting radio waves as they go. The radio jet in NGC 6251 is dramatically shown in radiograph form in Figure 4.1. The jet is a conduit along which energetic material carries energy and magnetic fields from the nucleus of the galaxy to the outer radio lobes. Figure 4.2 shows an overall view of NGC 6251, supporting the contention that the jet feeds the distant radio lobe. When the streaming matter hits the far boundary of the lobe, where the interface with surrounding intergalactic matter sets up an effective wall, a hot spot may form. There the jet splashes backwards and sideways and inflates the radio-emitting lobes.

A jet can be more than a million light-years long, such as the jet in NGC 6251 (Figure 4.1) which, at $1.2 \times 10^6$ light-years in length, is the straightest and longest known object in the universe. How are these jets formed and what holds them together? For the jets to be so long and straight a good "memory" is required, something which allows the flowing material to maintain a uniform direction for a very long time, a million years or more, which would be the travel time of some of the jets if the matter flowed at the speed of light. But how fast is that material flowing? Where does the energy come from which enables the jets to illuminate radio lobes a million light-years away?

Most of the extragalactic radio sources are double, but the jets themselves are usually one-sided. When two jets are seen they are usually very asymmetric as far as their brightness is concerned, as can be seen in Figure 4.2, a good example of this phenomenon. If jets are indeed conduits supplying the radio double, where is the other jet in one-sided jet sources, such as M87, shown in Figure 4.3? A radiograph of the jet of M87 is shown in Figure 4.4.

The answer may be related to the way these objects are oriented with respect to us; the jet streaming toward us may be much brighter than the one along which matter is receding; or, as one recent theory would have us believe, the jet alternately emerges from one side of the core of the radio galaxy and then the other, effectively flip-flopping back and forth over millions of years.

**FIGURE 4.1.** Radiograph of the jet in NGC 6251, a relatively nearby radio galaxy $2 \times 10^8$ light-years distant (compare with Figure 3.6). This extremely straight and narrow jet is about $1.5 \times 10^5$ light-years long. Small wavelike structures, indicative of instabilities (see text), begin to manifest themselves toward its tip. Vertical size is $5'$. (NRAO. Observers—R. A. Perley and J. Dreher.)

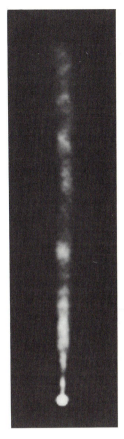

## The Energy Diet of a Jet

Jets are the channels along which power is supplied from a galactic nucleus into the extended radio source. The radio jet in M87 (Figure 4.4) is visible optically (Figure 2.3), but most radio sources have jets which are invisible but nevertheless are radio emitters which sometimes show associated X-ray emission as well. Jets are believed to carry equal amounts of electrons and positively charged ions out into the radio source, but only the electrons are observed, through the radio waves they emit. Although there is so far no direct observation of motion within the jets, the flow is estimated to travel between a few hundred and 10,000 km/s. The lower guess unfortunately leads to absurd conclusions about how old some huge radio sources must be—nearly as old as the universe—in order to inflate their radio lobes. The bulk motion of matter along the jet is far slower than the motion of the electrons, traveling near the speed of light, because these electrons (and ions) spiral around the fields and so do not progress very rapidly along them. When they do flow along the magnetic fields they

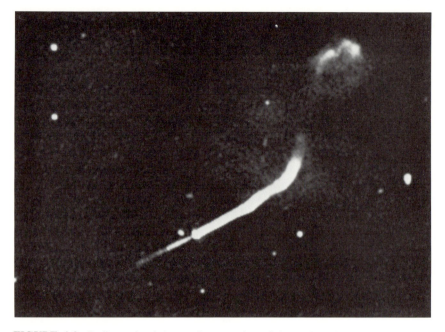

**FIGURE 4.2.** Radiograph of the northern section of the radio galaxy NGC 6251. In this image the other radio sources in the field of view were not removed during the data processing and lead to a more astronomical-looking image. The radio jet shown in Figure 4.1 corresponds to the bright part of this image, which is burnt out in this exposure. The parent galaxy is located at the left-hand end of this jet, and a very faint counterjet is seen to protrude toward the lower left. The much fainter emission in the outer lobes is difficult to see in this radiograph, and comparison with Figure 3.6 illustrates how the use of contour maps allows astronomers to study specific aspects of such radio sources. Vertical size is 25'. (NRAO. Observer—R. A. Perley.)

tend to bounce back and forth between irregularities in the magnetic fields, which can also trap the electrons into pockets within the moving clouds.

At the lower jet speeds it is possible to explain why some radio tails are smoothly bent. If the material were traveling very fast such smooth bending would be quickly destroyed. However, if the velocity is indeed low, several other problems are implied. For a radio source such as Cygnus A a mass ejection of 100 solar masses per year would be required in order to explain its luminosity. During a typical lifetime (around $10^8$ years—derived from the physics of the emitting region), the core would have had to eject $10^{10}$ solar masses into the radio components. This is more matter than is available in an elliptical galaxy; therefore, the velocity of matter in the jet cannot be so low. If, on the other hand, it is too high then the jets would not show the bends and kinks that they do. No one yet knows what the velocity in radio jets is, because whatever value is assumed leads to problems.

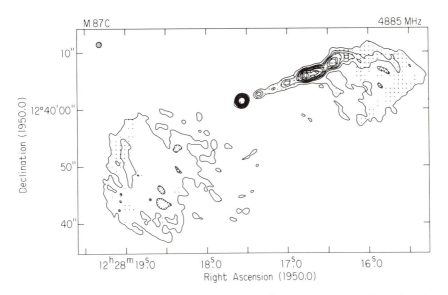

**FIGURE 4.3.** Contour map of M87, the $4 \times 10^7$-light-year-distant Virgo A radio source, made at 6-cm wavelength, showing that the radio emission is concentrated toward the core of the galaxy, which is marked by the dark ring of concentric contours. The radio jet extends to the upper right, is very bright, and feeds a lobe of low-level radio emission. Another, far more diffuse, lobe is evident at the left. (NRAO. Observer— F. N. Owen.)

The magnetic field structure of the jets, based on radio polarization observations, appears to be parallel to the jets in the more powerful sources (when a jet is observed) and perpendicular in the weaker sources, although even there they are parallel near the walls of the jet. (The field in the hot spots is usually aligned along the edges of the radio lobes, where the field may have been pushed up against the boundary.) This magnetic field in the jet probably helps keep the jet under control and stable over long periods of time; otherwise a flow from a jet would become unstable—that is, lose its ordered structure— and quickly destroy itself. This latter behavior is typical of jets observed in laboratory experiments.

The radiograph of the jet in M87, Figure 4.4, rather nicely shows the instabilities present toward the end of the jet where the object becomes more wavy in shape.

When radio doubles were first studied, it was suspected that something had to be flowing out from the galaxy in order to inflate the radio lobes. While no one specifically predicted that such narrow jets would be observed, and certainly not that they would be such beautifully organized radio-emitting structures, their discovery has turned out to be one of the most exciting topics in all of astronomy.

Jets have recently also been found on a much smaller scale. The very odd

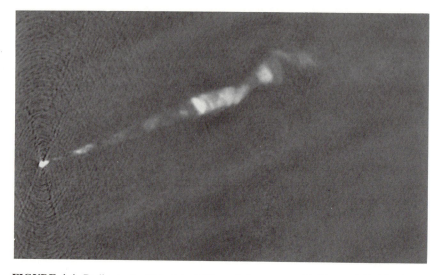

**FIGURE 4.4.** Radiograph of the jet of Virgo A (compare with Figure 4.3). The image processing on this map had not been completed when this print was made. Some 25 hours of computer time had already been used, but a further 200 hours of processing would have been required to remove the artifacts visible as faint rings in the radio image. The radio galaxy core is at the bright spot on the left. The structure of the jet, which is 4000 light-years long, shows striking "knots." Jet length is 15″. (NRAO. Observers—F. N. Owen, J. A. Biretta, and P. E. Hardee.)

object known as SS433, a spinning starlike object in our galaxy, emits two enormous jets (see Chapter 13), and motions can be observed within them, motions as fast as 78,000 km/s. Jets have also been observed in star-forming regions where they are known as bipolar flows (Chapter 10), and in galactic X-ray sources jets appear to be present on a stellar scale. Recently several stars in our galaxy have been found to have mysterious jets associated with them (Chapter 13).

Radio source jets usually contain a series of brightly emitting "knots." This is expected because the jet contains a magnetic field which sometimes gets kinks or knots in it. Matter will pile up at a knot and cause a brightening, due to increases in the local magnetic field strength and hence more synchrotron emission. The brightening indicates that more energy is being radiated away. This will cause cooling, just as anything that radiates away energy cools down, and so this region of the jet will contract, and possibly even collapse, as it gets cooler. This collapse increases the field strength and the emission. This cycle can continue indefinitely and creates an *instability* which can play havoc with the smooth flow in the jet. The magnetic kinks may be expected to move with the flow. Should anything get in the way, such as a cold cloud in the surrounding galaxy (and some have been observed near the core of Centaurus A), it will likely get swept up and accelerated until it is also part of the flow.

The region between the swept-up material and the flow, called a shock, may produce increased radiation, which can also explain the bright spots.

## Faster than Light—Superluminal Motions

As we move in close to the core of a radio galaxy, or a quasar, we often find something even more incredible. Radio-emitting material sometimes appears to be moving faster than light! This occurs not in the jet, but very close to the nucleus of the galaxy. There are about 20 objects whose bright compact cores have been studied with radio telescopes that span continents, and even stretch across the oceans (see Chapter 17). These extremely high-resolution radio telescopes reveal detailed structure down to fractions of a milliarcsecond in the cores of quasars and radio galaxies.

A stunning discovery about the bright radio cores is that they all show year-to-year movement on the smallest observable scales. Figure 4.5 shows a series of radiographs of the quasar 3C 345 made over a period of time. The systematic movement away from the core of a region of increased emission can easily be seen and corresponds to an apparent speed of between two and six times the speed of light. The maximum separation of the bright spots in this radiograph is 0.0009 arcseconds, equivalent to 12 light-years.

Since the distance to quasars and radio galaxies are known from their redshifts, it is easy to calculate how fast such blobs are moving outward. Depending on which source is being studied, they are found to move from 3 to 20 times the speed of light! This conclusion, however, is at odds with one of the best-known laws of physics—nothing can travel faster than light.

The discovery of this apparently superluminal (faster-than-light) motion in 1971 threw the astronomical community into temporary turmoil. Relative order was restored by the realization that one can get the illusion of superluminal motion through a peculiar projection effect. If a double radio source is pointed nearly at us, then we see the near-side material traveling toward us and the far-side material moving away. This has several consequences, some related to predictions made by relativity theory. Of course, the gas on the near side is blueshifted (moving toward us) and that at the far side is redshifted (moving away). However, if the ejection velocity is close to the speed of light the emission on our side becomes greatly intensified due to relativistic effects, whereas the far-side material would become so faint as to almost disappear. The apparent superluminal motion is a peculiar consequence of the fact that the material ejected toward us is traveling almost as fast as any light it emits toward us. A radio signal (traveling at the speed of light) cannot leave its source very far behind, and therefore two bursts of radio emission separated by a year could appear to us to be separated by a month, depending on the speed in the jet and on the angle between the jet and our line of vision. Therefore, when we see movement in the jets of the radio sources our initial estimate of the velocity of material could be completely wrong. This effect allows us to

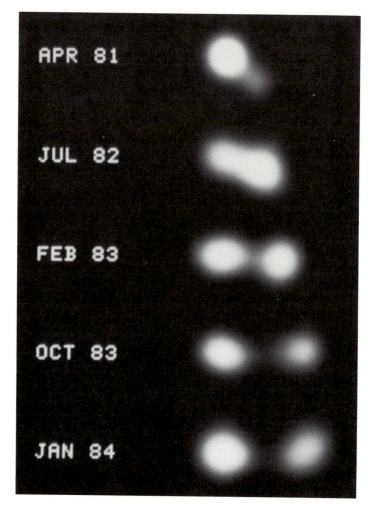

**FIGURE 4.5.** The "faster-than-light" phenomenon in the quasar 3C 345 revealed in a series of radiographs made over a three-year period. The emitting source on the left is stationary while the one on the right appears to move with a velocity from two to six times the speed of light. In the last map the separation between the components is 0.0009 arcseconds (12 light-years). (J. Biretta and R. Moore, California Institute of Technology.)

avoid the faster-than-light dilemma, but then another one pops up! If the larger, straight jets are related to the core jets, they too may be directed toward us, in which case they would be physically much longer than estimated. Therefore the jets may be far larger, and must be far longer-lived, than first suspected.

On the other hand, however, the intensity of the emission from a core jet pointed nearly at us is highly dependent on subtle relativity effects. The energy we think it is emitting may be far less than we originally believed! In fact, a quasar may be the extreme case of a core radio jet pointed directly at us. It would appear as a point source of radiation so bright as to dominate its parent galaxy.

Unfortunately, there is yet another problem implied by the existence of core radio source jets. Most of them show superluminal motions. We just decided that this can be explained if the jets are pointed nearly toward us, but why do they do this? Unless we occupy a very favored position in the universe, those distant jets should be pointed in random directions as seen from our vantage point. After all, the jets do not know we are here.

The solution may involve something scientists call a *selection effect,* which on occasion confuses many a research project. In this case we have specially selected those sources which have bright cores, including many quasars. Sources with bright cores may be just those that have jets pointed at us. That is what makes them bright and appear to be core sources in the first place! The ones where the jets are pointed across our line of vision would be too faint for us to see. So we expect to see superluminal motions in cores of radio galaxies and quasars. This is a controversial field of research, and more data which will help solve the mysteries will be forthcoming when the Very Long Baseline Array is built (Chapter 18).

Whether we solve the superluminal problem in the near future or not, it is obvious that at the very core of the sources very high-velocity flows are observed. But in the outer parts, in the larger-scale jets, the same material has slowed down to manageable proportions. Somewhere in between the matter has to slow down—to values of 5000–10,000 km/s, instead of those near the speed of light. This appears relatively certain because the properties of the outer jets are not consistent with bulk motions close to the speed of light (for example, the presence of smooth bends). At this time no one has successfully resolved these issues.

## Black Holes and Accretion Disks

So where does the flow of matter in the jets come from? What can eject two jets of material traveling nearly at the speed of light, and why would it continue to do this for millions of years? The answer comes from recognizing that the observed properties of the sources define the underlying nature of the hidden "engine" driving the radio source. Whatever it is, the engine shows two preferred directions, oppositely oriented in the sky. It is also very steady. Such a thing is a spinning object, acting like a gyroscope, which has two natural axes, its north and south poles of rotation—an object which can keep spinning very steadily for very long periods of time unless acted on by some external force tending to pull it out of alignment. The presence of two key directions, the south and north poles of the rotation axis, are natural directions along which matter could be ejected.

The invisible object is actually very small, but enormously massive. The only type of astronomical object that can satisfy the demands of the observations is a gigantic black hole. A black hole containing $10^7$ solar masses would be an object 3 light-minutes across, or approximately the size of Venus's orbit about the sun. A black hole containing $5 \times 10^9$ solar masses, such as is believed to exist at the centers of some radio galaxies, may be 28 light-hours across, or more than twice the size of the solar system.

A spinning black hole literally distorts the space around it, and any matter that comes relatively close will feel its tug, just as any particle feels the tug of gravitating objects in its neighborhood. For example, interstellar gas near the object will move inwards and will first settle into a disk which spins around the black hole. Interactions between particles of gas (known as viscosity effects) will force the particles to settle into this disk, and the same forces will cause the rapidly orbiting material to move gradually closer to the central hole. The gas will grow hotter as more and more energy is released due to collisions and interactions in the swirling motion about the black hole. This gas will grow so hot, as much as a billion degrees K, that it will actually expand and form a fat torus—a doughnut-shaped region—rapidly spinning around the black hole. This is illustrated in Figure 4.6. Inside this torus will be magnetic fields which are literally tied to the black hole because some of the gas will have plunged into the hole and dragged the fields with it. Those magnetic fields at

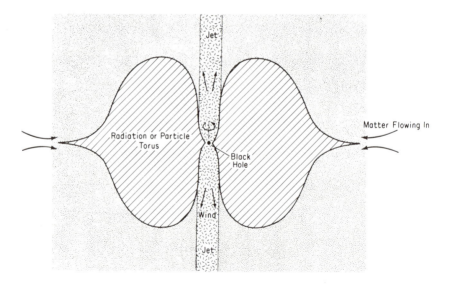

**FIGURE 4.6.** Model for the region around a black hole at the nucleus of a radio galaxy or quasar. Matter flows in and fills a torus (doughnut-shaped region). Material that leaks into the central funnel will be driven outward at velocities approaching the speed of light and will form two jets of material which subsequently power the lobes of a double radio source such as Cygnus A (Figure 3.5).

first remain connected to the gas outside and will rapidly wind up. Then, when they have become over-wound, they will snap. As a result the fields will realign themselves, but since they are constantly being wound up they will snap again, and in this way energy from the rotating black hole is converted, through the magnetic field reconnection process, into the energy of particles where the field is being so badly twisted and distorted. This process continues as long as the appetite of the black hole and the availability of gas allow it. Matter keeps falling in and at the same time energy from the black hole escapes as field lines are broken. The gas grows even hotter.

Around the poles of the black hole there is a critical funnel-shaped region of space (the narrow hole of a fat doughnut; see Figure 4.6) in which matter finds it has two options. If it has insufficient energy, it will plunge into the black hole and wave the universe good-bye. If, however, a particle has enough energy, it may suddenly be free to escape from the funnel and to blast out into space!

If the shape of this funnel is narrow enough, provided the torus of gas around the black hole is thick enough, this matter will escape as if ejected from a nozzle. This may well produce the jets seen in the cores of radio galaxies and quasars. No other plausible explanation has been suggested.

A similar beam of high-energy particles leaving the nucleus of a spiral galaxy would tend to collide with surrounding interstellar gas, so abundant in spiral galaxies, and this gas would obstruct the flow. Hence black holes at the centers of spiral galaxies would not be likely to create jetlike radio sources. Only in relatively gas- and dust-free elliptical galaxies will the gas stream outward and be likely to escape unimpeded, as is observed.

The energy of this ejected material comes from the black hole itself. Based on an efficiency of 10% for the energy creation process, the most active galactic nuclei in distant radio galaxies must have processed $10^8$ solar masses through a region not much larger than the solar system. Since the sources are believed to have lifetimes of about $10^8$ years, that requires only one solar mass to be processed per year and only a fraction of it escapes up the funnel. That is enough to cause the radio galaxy to glow. The black hole has to be about $10^8$ solar masses to do the job of creating a double radio source.

Because a spinning black hole is an excellent gyroscope, it will continue to point in the same direction in space as long as is demanded by the astronomical observations. Some black holes do not maintain such a nice peaceful existence, however.

## Precession

The jets of some radio sources show a wavy jet structure, almost as if the ejection occurred out of a hose which was being swung around in a circle. It is believed that this is precisely what would be expected if the black hole is *precessing,* that is, if its axis is slowly wobbling about its mean position in

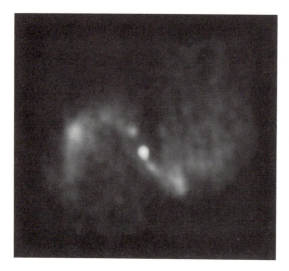

**FIGURE 4.7.** Radiograph of the precessing jets in the radio source 4C 29.47 observed at 20-cm wavelength. The radio source is associated with an elliptical galaxy $6 \times 10^8$ light-years away. The brighter patches in each jet signify the occurrence of periodic out-bursts which simultaneously inject material into both jets. Vertical size is 3'. (NRAO. Observer—J. J. Condon.)

the sky. The earth's axis also precesses. It does not always point the way it does now but precesses slowly, making one cycle every 26,000 years. This precession is due to the tug on the earth's axis from two nearby gravitationally attracting objects, the sun and the moon. A top or a toy gyroscope spinning on a table top also precesses (wobbles) because gravity is tugging at it.

A black hole may precess because of a gravitational pull from another object, perhaps another black hole, a galaxy, or a dead quasar passing close by. The flailing jets of 3C 75 (Figure 1.3) may be caused by the precession of two interacting black holes. Another possible example of precessing radio jets is shown in Figure 4.7, which is the radio galaxy 4C 29.47. In Figure 4.8, which shows quasar 4C 18.68, a theoretical calculation of a precessing jet has been compared with the radio data and shows fair agreement, suggesting that the model may be correct. The precession is nevertheless extremely slow, taking many millions of years to complete one wobble.

## Galactic Cannibalism

Given that we have a black hole doing all this wondrous stuff, we have to ask where the gas comes from that continually fuels the ejection from the black hole. That is a story that involves galactic cannibalism.

When galaxies are very close together, as in the early universe or in dense clusters of galaxies in which radio galaxies are found, they may sometimes wander into each other's gravitational pull and become linked in an embrace that inevitably lures one galaxy into the other's cooking pot. We imagine one galaxy as a cannibal and a smaller galaxy, the missionary, tentatively, but inexorably, approaching it. If the missionary gets too close, it will get stripped

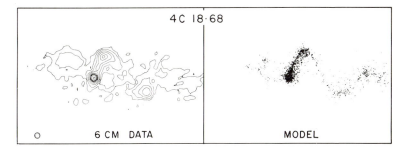

**FIGURE 4.8.** 4C 18.68, a quasar with precessing radio jets. The left-hand diagram is a contour map of the radio emission from this quasar, which lies at the position of the brightest peak. The right-hand image is a computer simulation of what would be expected if two precessing jets were spinning out from the quasar; a larger number of points indicates where a greater intensity of radio emission is expected. This model agrees well with the observations. (Ann C. Gower, University of Victoria.)


of its gases. Then the missionary finally falls into the cannibal's caldron that is the nucleus of the galaxy. This cosmic feast provides vast amounts of material which can be gathered into the accretion disk around the cannibal's black hole. The fatal embrace between the two galaxies is long and slow, just what is needed to continually feed the black hole a steady diet of fresh gas. Jets are then blasted into space through the funnel around the black hole, allowing a reprieve for a small percentage of the captured particles.

Calculations have shown that gas and dust torn from a nearby galaxy could have settled into the shape observed in NGC 5128 (Figure 3.4). It is also significant that the clusters containing the greatest numbers of galaxies (rich clusters) often have the most powerful radio sources near their centers. These are regions of overcrowding, just where cannibalism is most likely to occur. It is here that tidal encounters (close encounters in which the gravitational pull of one object distorts the other) are more likely, and these create frictional forces which cause galaxies to spiral inward until they are close enough to gobble each other up. The most massive ones would form huge star densities at their cores, star clouds which would provide an endless and systematic repast for the gourmet black hole. NGC 5128 is regarded as manifesting a relatively mild form of cannibalism which may have been more common in the early universe, when galaxies were more closely packed (the universe was smaller then).

Stars themselves may be involved occasionally. Old stars, as well as exploding stars, may provide an alternative source of material to feed to the black hole. The brightening of the cores, the optical variability, and the ejection of blobs of material along the jet in the form of superluminal motions may be related to sudden bursts of matter consumption on the part of the black hole, perhaps a sign that a star or two went down the wrong way.

The original formation of the black hole is also a mystery. How did $10^8$

solar masses of material come together? Did such enormous entities form in the primeval universe, in one fell swoop, in some gigantic, unimaginable cataclysm, or did smaller black holes, formed by exploding stars, inexorably consume each other until they coalesced into the enormous trap they have now created? More intimate probing of the cores of radio galaxies and quasars is needed to answer these questions.

# 5

# Radio Galaxies and Quasars: An Overall View

## Energy Supply in Radio Sources

The recent radio observations of the detailed structure of the radio sources and their jets have led to a genuine advance in knowledge about these objects. They appear to be fueled by twin jets that emerge from the region of a spinning black hole located at the centers of the parent galaxies. (Further examples of radiographs of radio source jets are shown in Figures 5.1 to 5.4.) However, there are two continuing problems associated with understanding the fueling of the radio source: First, how do the jets and the radio lobes, located so far from the central "engine," the black hole, continue to emit radio waves for tens of millions of years? Because the matter moves relatively slowly along the jets and has to travel up to a million light-years distance, we know that the particles have to keep radiating energy for that length of time. But there is no way to feed the particle enough energy at the outset of its journey and expect it to keep shining for so long. The very process of emitting radio energy means the particle loses energy and it should cease to shine. Secondly, sources like Cygnus A are known to emit far more energy from their outer lobes than is being ejected from their cores by even a hypothetical explosive event involving millions of stars. Obviously the jets must obtain a transfusion during their journey, probably several transfusions, which will allow the radio lobes to be radio bright even after tens of millions of years.

Radio astronomers have been confronted with these two problems ever since the double radio sources were first discovered; how can the radio sources emit so much energy, and how can they do so for long periods of time? A fascinating new explanation has recently been proposed. The jets do indeed obtain energy by transfusion, transfusions of the most remarkable kind.

Bulk kinetic energy (energy of motion of large masses of material in the jet) can be converted into particle acceleration in the jet through the action of turbulence (that is, chaotic motion) within the jet itself. Much of the energy may be generated within the jets as well as in the radio lobes and may not have originated near the central black hole after all. As the jets blast through space, they draw energy from surrounding interstellar and intergalactic gases.

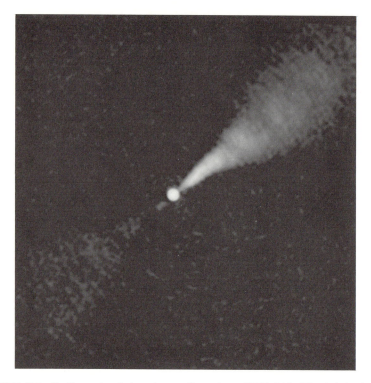

**FIGURE 5.1.** Radiograph of the giant radio galaxy NGC 315. The bright jet and fainter counterjet are enormously long structures, reaching $2 \times 10^5$ light-years from the central galaxy. Vertical size is $4'$. (NRAO. Observer—A. H. Bridle.)

To account for this, at least three mechanisms for adding energy to the relativistic particles have been proposed: shocks, turbulence, and entrainment.

*Shocks,* which are sudden discontinuities in the properties of the material involved, such as abrupt changes in density or motion, ahead of the jet and along its walls, create strong magnetic fields which can accelerate particles. This process converts energy of bulk flow in the jet into relativistic energy of electrons; that is, the electrons are accelerated close to the speed of light. Bear in mind that the particles, after radiating their energy, slow down to become nonrelativistic.

*Turbulence* in the medium can create a similar acceleration. Electrons may also be reenergized by collisions with protons in the stream. Thus we have particles continually propelled back to close to the speed of light.

As the material streams outward, escaping through the funnels in the torus near the black hole (Figure 4.6), matter is dragged or sucked in from the medium surrounding the jet. This process, called *entrainment,* has very interesting consequences. Whirlpools of matter, known as eddies, will be set up around the edge of a jet, just as in the wake of a ship. These eddies create shocks. The shocks can heat the gas to $10^7$ K, and some of this gas then accumulates

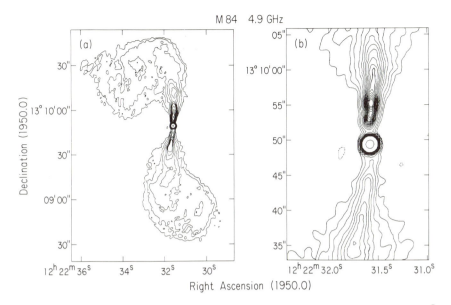

**FIGURE 5.2.** Radio contour map of the radio galaxy M84 located about $2.7 \times 10^7$ light-years away. The right-hand panel (shown in radiograph form in Figure 5.3) reveals details of the central region—the parent galaxy is located at the circular contours. Two jets feed energy into the extended lobes and the northern jet is clearly brighter than the southern one. This asymmetry is typical of many radio sources. (NRAO. Observers— R. A. Laing and A. H. Bridle.)

into dense pockets behind the shock. This entrained gas is predicted to cool to 10,000 K and show optical emission lines, just as observed in Centaurus A. By this time the gas will have moved as far as 30,000–300,000 light-years from the core. It is postulated that some of this gas will cool further, and stars will form. This has also been observed in Centaurus A. These stars will subsequently evolve, age, and die as they move out with the jet. Many will end their lives in explosive deaths known as supernovae (see Chapter 8) and these explosions will, amazingly, become a significant source of new energy for the radio jets and lobes. Bear in mind that the material in the jets takes millions of years to reach the radio lobes and therefore stars have plenty of time to be born and die during their journeys along these conduits into space. Entrainment, therefore, leads to a series of events, including star formation, which keeps the jet refueled and the radio source ''shining.''

## Other Radio-Luminous Objects—Blazars and Seyferts

The radio galaxy and quasar story has become dramatically clearer with the discovery of several other classes of galaxy which exhibit violence in their cores. For example, BL Lac objects are variable starlike objects discovered in

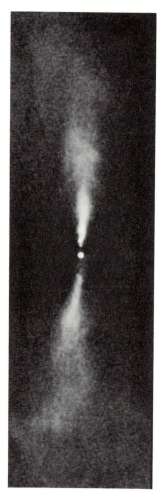

**FIGURE 5.3.** Radiograph of the central regions of the radio galaxy M84, equivalent to the right-hand side of Figure 5.1. M84 is located in the Virgo cluster of galaxies, as is M87 (Figure 4.3). Vertical size is 2'. (NRAO. Observer—A. H. Bridle.)

1929. (The abbreviation, Lac, refers to the constellation Lacertae in which the first one discovered is located.) These objects turned out not to be stars at all, but nearby elliptical galaxies, many of which are radio sources. Their unique characteristic is their extremely short-term optical variability, their brightness changing by a factor of up to 100 (or the light of $10^{10}$ suns) in a few months. It is as if $10^{11}$ stars turn on and off together. The prototype of the class, BL Lac, exhibits a tenfold change in its brightness in a few months. Sometimes changes are observed from day to day.

**FIGURE 5.4.** Radiograph of the jets in the radio galaxy 3C 31, $5 \times 10^7$ light-years distant. Both jets widen as they leave the nucleus of the galaxy. As is typical for many two-sided jets, one side is brighter than the other. Vertical size is 2'. (NRAO. Observer—A. H. Bridle.)

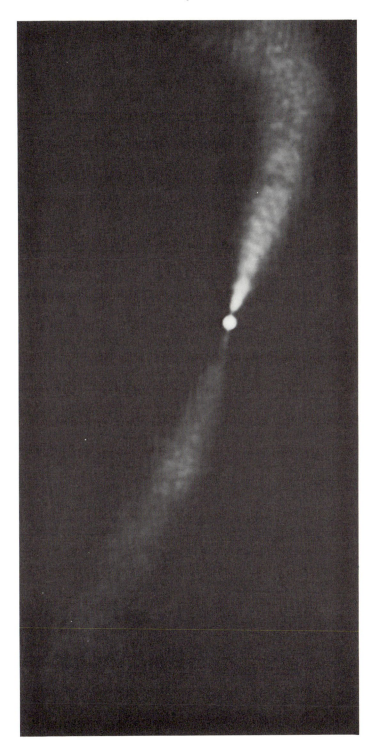

Radio outbursts from BL Lac objects are also very intense, but apparently uncorrelated with the light changes. These objects are sometimes called *blazars* and appear to be very similar to quasars. In the blazars we may be looking straight down on top of the central "engine" in the nucleus of the galaxy. When the blazar is quiet it looks like a quasar, with its parent galaxy visible. When it suffers an outburst, which generates great heat, neither spectral lines—characteristic of cool matter surrounding quasars—nor evidence of the parent galaxy can be seen, due to the glare.

Another class of peculiar spiral galaxies, called *Seyfert* galaxies, are relatively faint radio sources. They are peculiar because they show broad optical emission lines at their cores, the broadness indicating extremely chaotic, violent movement at the centers of these galaxies, motions which are not found in the more peaceful and normal spiral galaxies such as our own. Many Seyferts also show peculiar and distorted appearances on optical photographs. Their radio emission is usually confined to the core region, and if a jet is present it is usually very short, less than a thousand light-years long. Galactic cannibalism may be a reason for intense activity in these galaxies. Seyferts may be a class of objects between radio galaxies and quasars.

## Review of Radio Source Structures

We now summarize what is known about the extragalactic radio sources and paint a picture of how they may be related.

Radio galaxies are the brightest and largest members of clusters of galaxies. Quasars are also likely to be members of clusters of galaxies, but because of their great brightness and huge distances, the surrounding cluster members are relatively invisible to us in all but a few cases. The basic types of radio source are:

1. Narrow, edge-brightened double sources. Tails may reach from the center toward hot spots. The quasar 3C 273 is a one-sided source related to these.
2. Narrow, edge-darkened double sources, such as Centaurus A (Figure 3.3) and 3C 449 (Figure 3.7).
3. Wide double sources, such as 3C 310 (Figure 3.2).
4. Narrow-tailed sources, such as NGC 1265 (Figure 3.8). They have been described as tadpole-like, are asymmetric, and are also known as head–tail sources.
5. Wide-tailed sources, such as M84 (Figures 5.2 and 5.3).

The overall shape of radio sources is defined by several characteristics, including the degree of bending, which ranges from slightly bent to completely swept back. The sources with a greater bend have a larger density of cluster galaxies around them, and sources inside clusters are more complex than those located outside. Another important characteristic is their symmetry. For many sources we can find an imaginary axis about which to rotate the radio maps so that the two halves of the source mirror each other.

The orientation of radio source doubles with respect to the optical structure in the parent galaxy is striking. There are at least seven unambiguous cases where the double radio source is oriented precisely perpendicular to the dominant (major) axis of the galaxy as defined, for example, by a dust lane such as the one in Centaurus A.

## Grand Unified Scheme—From Quasar to Milky Way

An "engine" imbedded in the core of a galaxy ejects a narrow stream of material in two opposite directions along its rotation axis in the form of a continuous blast of relativistic particles which spiral about magnetic fields as they move outward. This causes the particles to emit radio waves by the synchrotron process. As a direct manifestation of such core activity, we often see radio and optical (quasarlike) sources in the nuclei of these galaxies. If the ejection continues long enough and is stable, extended radio lobes are created. The shape of the source depends on the degree of activity in the nucleus and on the nature of the surrounding cluster medium, as well as the rotational motions of the source with respect to this medium. Whether a quasar or radio galaxy is formed depends on how long ago the process began. In the early universe quasars predominated. The probability of an individual galaxy becoming a radio source or quasar appears to have been 1000 times greater in the early universe than more recently. Quasars appear to be galaxies with extraordinarily luminous nuclei.

In quasars the central black hole may be so massive that most of the matter near it is prevented from escaping into surrounding space and the core is the primary emitter of radio and light energy. On the other hand, a core jet may be moving almost directly toward us, giving the illusion that it is a central source.

Closer to our own epoch, in more recent times, the radio galaxies predominate. More matter escapes out of the funnels near black holes and the jets then feed the enormous radio-emitting lobes (for example, the radio source shown in Figure 5.4 may be compared with the model of the black hole region shown in Figure 4.6).

In the early universe, when fuel was plentiful, most sources were quasars. In elliptical galaxies, where stars are plentiful and nearby galaxies are available to be readily consumed, the quasars phenomenon is radio loud. But in spiral galaxies (related to the Seyferts), with fewer stars near the nucleus, and in sparser volumes of space, the black holes find less fuel and the quasar phenomenon at the core may be radio quiet. According to recent discoveries, the radio-quiet quasars (QSOs) are actually in the majority, constituting 90–95% of the sample.

As fuel runs out in an elliptical galaxy, the optical activity would diminish, turning the galaxy from a quasar into a strong radio galaxy and later into a weaker one. Black holes may exist in the cores of all galaxies. Within elliptical galaxies there may be black holes containing $10^7$ or even $10^8$ solar masses while those in spiral galaxies may only contain $10^6$ solar masses or less. The nuclei of spiral galaxies, such as the Seyferts, with their smaller black holes, might maintain their optical activity for longer periods of time. They would subsequently weaken into relatively quiet galaxies like our own. The Milky Way may once have been a Seyfert and, indeed, a $10^6$-solar-mass black hole may reside at the center of our galaxy.

# Part III

## The Milky Way

# 6

# The Galactic Center

## The Milky Way Galaxy

The sun, its attendant planets, and all life we know of drift 27,600 light-years from the hub of our Milky Way galaxy. (This distance was settled on by international agreement in 1985, although a value of 30,000 light-years has been the favored one for several decades.) Seen from our vantage point, the stars and matter between the stars lie along a great band across the sky (Figure 6.1). The earth and sun are located in a quiet region of space, that is, quiet compared with the violence we have witnessed on our journey to the centers of radio galaxies. Yet the center of our own island of stars is itself a fascinating region of chaotic activity and mystery, whose secrets are revealed only when we open our eyes to the invisible radiations of the radio and infrared.

The Milky Way is the name sometimes given to our galaxy, a vast, disklike distribution of stars, gas, and dust, at least 100,000 light-years in diameter and 1000 light-years thick. The sun is located inside this disk. The Milky Way is also the name given to the band of faint light which crosses the night sky. The glow is produced by the light of millions of distant stars, too far away to be seen individually. Away from the faint glowing band we are looking at right angles to the flat plane of the Milky Way, and fewer stars are present. Here we are able to look farther out into space where most of the distant galaxies are visible. On the other hand, individual stars that make up the constellations are relatively close to us, usually between 10 and 100 or so light-years away.

## The Galactic Center

One of the most fascinating detective stories in radio astronomy concerns the remarkable progress made in understanding the mystery of the galactic center. Its light is completely obscured by dust; less than a trillionth of the light from stars near the center can reach us. Astronomical studies have therefore focused on radio and infrared observations. Because their wavelengths are much larger

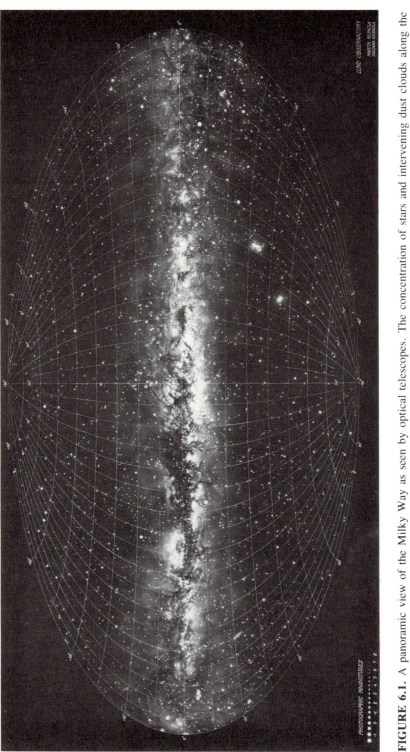

**FIGURE 6.1.** A panoramic view of the Milky Way as seen by optical telescopes. The concentration of stars and intervening dust clouds along the Milky Way, also known as the galactic plane, is dramatically obvious. The dust clouds toward the center of this panorama obscure the center of the galaxy in the direction of Sagittarius. (Lund Observatory.)

than the size of the dust particles that drift in great clouds between the stars, these waves travel relatively unhindered through the dust clouds. Light, on the other hand, has a wavelength about the same size as a dust particle, and is therefore mostly absorbed. The enormous recent technological progress in astronomical telescope construction and instrumentation has been exploited to study the galactic center, and every observation with better equipment has produced stunning new information. The subsequent detective work has been equally stunning in its complexity, with our knowledge of the physics and chemistry of interstellar matter pushed to new limits in order to form the pictures described below.

## The Radio Source in Sagittarius

The very first cosmic radio signals to be discovered, by Karl Jansky, appeared to be associated with the Milky Way band of stars (Figure 6.1), and the greatest intensity was located in the direction of the constellation of Sagittarius. Grote Reber (Chapter 16) made follow-up measurements with a radio telescope specifically designed for the task, and confirmed that the peak of the emission lay in that constellation. In 1959 the central radio source was recognized as being composed of at least four separate sources, labeled Sgr A, B, B2, and C. Sagittarius A (Sgr A) is the generic name given to the very complex structure revealed in the direction of the galactic center. (For the observers of the night sky this strong radio source is situated almost at the boundary between the constellations of Sagittarius and Ophiucus, and lies about 4 degrees beyond the tip of the spout of the ''teapot'' which defines the visible constellation of Sagittarius.)

Until the 1950s it was believed that the center of the galaxy was located in a direction about 30 degrees away from Sgr A, such a misidentification was to be expected in view of the difficulty of seeing anything in those directions with optical telescopes. Only with the advent of radio astronomy did a clearer picture of our galaxy become possible. It was the discovery of this strong radio source that began a trail of research which led to the present definition of where the center of the galaxy must be located. Within the Sgr A complex is a bright, compact, pointlike source of radio waves, labeled Sgr A*, which may be similar to the compact nuclei of radio galaxies. This source is suspected of being as close to the absolute center of the galaxy as anything known. [Again, for the serious observer, its location is at right ascension 17h 42m 29.3s and declination $-28°$ 59' 18'' (1950) and, by definition, galactic longitude zero and galactic latitude zero.]

## Radio Images of Sagittarius A

A beautiful radiograph of the galactic center region is shown in Figure 6.2. This image was the result of hundreds of hours of observations and an equal amount of computer time afterwards. The distinctly different sources Sgr A,

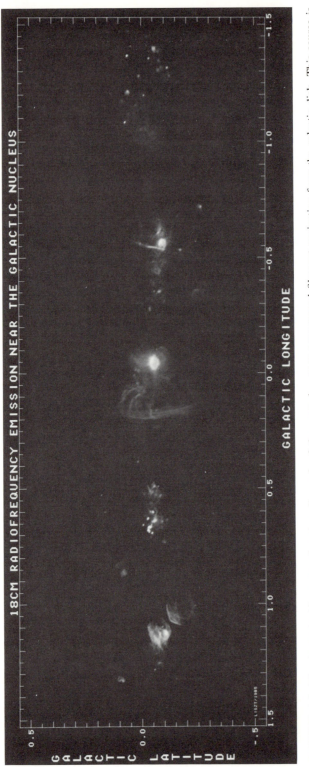

**FIGURE 6.2.** The dramatic 18-cm-wavelength radiograph of the central region of the Milky Way. Sagittarius A, located at the very heart of the Milky Way, dominates the map's center—this section is shown in detail in Figures 1.5 and 6.3. Sagittarius C is the source to the right, showing a very pronounced filament projecting from the galactic disk. This source is believed to be about 300 light-years from the center. (NRAO. Observer—H. S. Liszt.)

B, B2, and C, as well as Sgr D, are obvious and each of them is unique in appearance. Figure 6.3 is a series of radio images and maps which zoom in on the detailed structure in Sgr A and serve to illustrate the remarkable advances made in radio astronomical observations over the last two decades.

The early single-dish observations (upper right diagram in Figure 6.3) showed a broad band of emission, with several peaks superimposed, running along the galactic equator. The most intense of the peaks is Sgr A, and there was a suggestion of an arc in its structure. The radiograph of this central region, shown in the upper left frame and in Figure 1.5, illustrates the extraordinary resolution attained by the Very Large Array (Chapter 17). Both straight and curved filaments of radio-luminous material arch away from the nuclear region itself (the bright blob of emission within which Sgr A* is buried). The straight filaments, possibly encircled by looped structures, cross the galactic equator at right angles; they are shown in more detail in Figure 6.4. (Note that one arcminute on these maps corresponds to a linear dimension of 10 light-years at the distance of the galactic center.)

The next image, at the bottom left of Figure 6.3, is a close-up of the Sgr A complex which shows a ringlike structure centered below the galactic equator. The bright Sgr A* source appears as a series of closed contours located inside an S-shaped region which is further blown up in the bottom right frame. An extraordinary display of spiral-like features now becomes apparent, and although reminiscent of a miniature spiral galaxy, it is believed, on the basis of the Doppler shifts of spectral lines from these features, that there is a patchy, ringlike configuration of gas rotating about the center, and possibly infalling motion along the bar crossing the center of the ring.

The entire galactic center region may be far more complex than ever suspected. For example, a wider panorama of the view toward the center, based on observations made by the new high-resolution radio telescope in Japan, shows a protuberance jutting from the Milky Way (Figure 6.5) above Sgr A as well as one a few degrees from it, both of which have an appearance reminiscent of the ubiquitous radio jets in the nuclei of distant radio galaxies. Whether or not these protuberances are indeed indicative of great activity in the galactic nucleus remains to be determined.

## Journey to the Center of the Galaxy

Imagination is an important tool in science, and it is particularly important on our journey through the invisible universe. Therefore, before we complete this chapter with a detailed description of what the radio observations reveal, let us take an imaginary journey to the center, taking in all the sights as we go. To make the voyage more dramatic, let us first lift ourselves out of the plane of the Milky Way (Figure 6.1), out of the disk of stars, and travel up and out, until the sweep of the galaxy, from its heart to its outer realms, fills our vision. Imagine that the galaxy looks something like the spiral galaxy shown

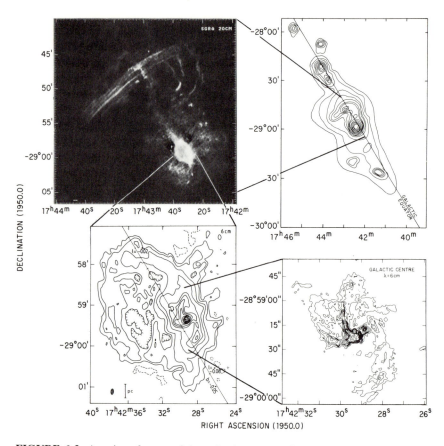

**FIGURE 6.3.** A series of maps of the galactic-center radio source. The top right-hand figure is an early-1960s map of the radio emission for the entire region, whose extent may be directly compared with Figure 6.2. The upper left image is a radiograph of Sgr A, which is seen at the center of Figure 6.2 and in the close-up in Figure 1.5. The remarkable increase in resolution between the upper two frames is striking. The existence of the filaments was already suggested in the older data. The lower left diagram is a more detailed view of the core source and shows that it consists of a ring which, in turn, contains the spiral-like structure (lower right). The "dead-center" of the galaxy, the Sagittarius A radio source, lies at the center of this map. Note that the coordinates are right ascension and declination and the galactic plane crosses these maps as indicated. [Observers: (a) F. Yusef-Zadeh, M. R. Morris, and D. R. Chance; (b) D. Downes, A. E. Maxwell, and M. L. Meeks; (c) R. Ekers, J. H. van Gorkum, U. J. Schwarz and W. M. Goss; (d) K. Y. Lo and M. J. Claussen. Reproduced, with permission, from *Nature*, Vol. 306, p. 649. Copyright © 1983, Macmillan Journals Limited.)

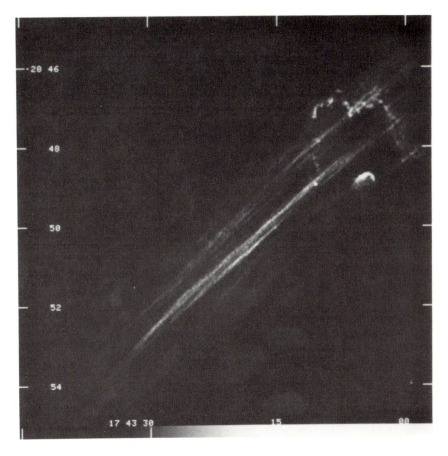

**FIGURE 6.4.** A detailed look at the filaments in the arc associated with the galactic center. The straight filaments break up into many narrow structures several light-years in width. Another filament seems to twist around these. (NRAO. Observers—F. Yusef-Zadeh, M. R. Morris, and D. R. Chance.)

in the Appendix (Figure A.1). Imagine your eyes encompassing more and more of the scene, as you drift out there in the darkness of space.

Now behold the fiery Milky Way with its reds, greens, blues, violets, pinks, and intense points of white light, lying before you. Billions of stars shine in a great glow that spreads across your vision, billions of stars interlaced with bright bands and dark regions where spiral arms weave through the galaxy. Over to your left the swath of stars fades toward the galaxy's boundary with intergalactic space. Over to the right a bright glare marks the vast spherical star fortress within which lurks the center of the galaxy. Behind you the darkness of empty space is marked only by patches of light from a hundred clusters of stars wandering in perpetual orbit around the galaxy. Perhaps you will notice

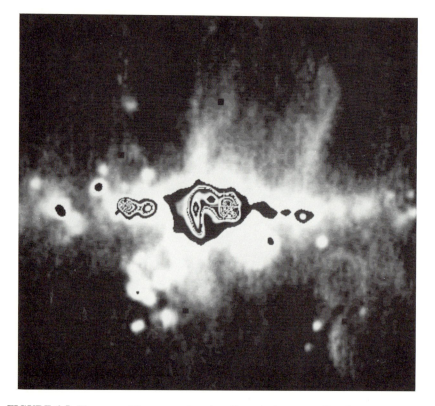

**FIGURE 6.5.** The spur of 3-cm-wavelength radio emission protruding from the galactic center region as mapped with the Nobeyama Radio Telescope in Japan. Overlaid on this radiograph are the main features of the galactic-center radio emission recognizable from the maps in Figures 6.2 and 6.3. The orientation of the radiograph allows direct comparison with Figures 6.1 and 6.2. (Y. Sofue and T. Handa, Nobeyama Observatory.)

light from distant galaxies, fuzzy patches, some of which may even show spiral structure to your new eyes.

Travel now, travel toward the hub that beckons in the distance. Sweep low over the disk of the galaxy and watch the bright swaths of dust and star-filled spiral arms surge by beneath you. They are 5000 light-years from side to side and 1000 light-years thick. You can see into and along the dark canyons between the arms, and perhaps, as you hurtle on, you might notice a distant galaxy or the Clouds of Magellan, our galaxy's nearest neighbors in space.

Approach the vast swarm of stars that so vividly delineates the central regions of the Milky Way, the hub about which 250 thousand million stars ponderously orbit, stars with which we share our common island in space. Downward you zoom, toward the teeming mass of matter, toward the dark clouds which drift between the stars. Soon you are inside and the light from the stars grows dim and tired for it cannot push past the dust that drifts between the stars.

In darkness once again, you pause to adjust the controls on your cosmic eyes so that they will be sensitive to infrared radiation, the heat from stars and gas and dust, which can pass undimmed by the obscuration of dust particles. The universe about you opens up and you can see again. The obscuring dust has grown transparent and a red glow from thousands of distant stars draws you toward the nucleus. Move on through the haze and the glow ahead grows brighter while around you it is still dark. Suddenly you burst into a great cathedral in space where you stop, four light-years from the center. Look around, take in the heat and the light and sense the architecture of the galactic nucleus.

Nebulous clouds drift everywhere; multicolored tones splashed amongst the vivid reds, blues, and whites of starlight. Dense clouds of dust, their outer boundaries cocoons of fiery heat, drift across the field of view. Some are silhouetted against a great cluster of a thousand reddish stars which drift in the distance.

Now tune your eyes further. Allow the radio signals in. Sense their energy. Ethereal structures are revealed through the radio waves they emit. Your radio vision clears, the stars disappear and a new panorama unveils itself. An enormously bright center of radiation glows in the distance, sometimes flaring, but always asserting its dominance over all the matter and energy that exist here. All around the nucleus great patches of enhanced emission spiral and swirl, forever subject to the assertions of the great mass at the center.

You begin to sense that the radio-luminous forms define a great circular wall of glowing matter that circles all around. You are near a segment of this wall, and matter at the surface seems to be organized in great clumps, some of which have detached themselves and are glowing as separate entities. All around, as your eyes adjust, you sense a glow that fills the apparent void encompassed by the great ring of matter. A swath of luminous material cuts across the ring in the distance, and behind you, beyond the luminous wall of the ring, you sense a darkness looming where great clouds of molecules radiate their signatures for all to "see." And beyond them, long filaments of emission, streamers of luminosity arching up and out of the nucleus, trail into the distance 20 or 30 light-years away.

Let all the radiations in, the radio, the heat, the light, and the X-rays. The heavens around you are radiant. A sudden vivid flash of gamma rays blinds you momentarily. Then your attention returns to the great source of radio waves known on earth as Sagittarius A*, a cluster of fuzzy red stars, each sheltered in a cocoon of glowing gases.

Next, adjust your perception of time. Allow time to hasten and watch the movement all about. Clouds swirl and heave and tumble. Stars rush by, swaying about, oscillating up and down as they go, surging this way and that, pulled and pushed. The ring of matter is revolving, rotating, moving in a great circle. Its gases churn and condense to tear themselves free and plunge through the glow toward Sagittarius A* and the cluster of red stars that are locked in an intimate embrace. The cluster itself pulses, and stars boil toward the surface only to be pulled back by hidden forces of attraction.

Watch as stars brighten and fade, and new ones burst upon the scene. Watch as another glows and brightens, then suddenly explodes in a flash which lights up the surrounding swirling clouds of dust. Watch as the shell of luminous matter tears outward and is tugged this way and that by surrounding gases. Watch the shell of stellar debris distort as it travels outward and fade as it returns its remnants to the interstellar dust from which all stars were born.

Listen to the vibrations emitted by chemical and physical changes in the surrounding gases. Listen to the rhythm of the atoms and molecules, as electrons leap in their orbits, atoms vibrate to the rhythms of their internal forces. Listen to the symphony of sound that resonates within this cathedral of space. You are witness to primitive cosmic evolution. You can almost hear life surge through this vibrant volume in which a secret, massive force holds everything together.

Still invisible to all your senses, hidden deeply within the clouds of fiery gas, sheltered behind stars swaying in ordit about its unseen presence, a massive black hole embraces its cohort and draws matter inexorably into its grasp. Radio-emitting electrons tumble inward, gathering in a concentrated disk of swirling gas, luminous and boiling with energy, locked in great energy, locked in a peculiar death dance, drawn toward the black hole, drawn in to add to its awesome mass.

The black hole is invisible, yet you sense its dominance. You watch the hot gases move almost to its edge. There the glow becomes redder and redder before it fades into an invisible boundary and is gone forever.

All about you, subtle patterns or energy and movement speed amongst the stars and gas, waves running this way and that, reflecting unseen forces, interplaying, intermingling, determining the fate of the gas and the dust and the stars. The movement is wondrous. This is the heart of the galaxy, a relatively normal, ordinary place, where events are more orderly than in distant radio galaxies or quasars, and peaceful enough to have allowed life to emerge on at least one distant planet.

Now move into the very center, into the cloud of stars, into the radio aura. You are at the hub, at the very center of rotation, of the Milky Way. Perhaps you can sense its rotation, like the potter feels the center of the mound of clay spinning on the wheel, sense the formless axis which allows your touch to enter smoothly and where all potential is felt. You are the galaxy now. You feel the motion. You sense the dynamic play of the cosmic cycle of birth, life, and death. Is the heart of the galaxy alive?

## The Galactic Center in Close-Up

To conclude this chapter, a more traditional summary picture of the galactic center is presented, a view based on a distillation of an enormous amount of work by very many people. While details are often controversial, this broad summary is probably a fair reflection of current consensus amongst astronomers. A tremendous amount of research continues to focus on this interesting region.

The Sgr A source is a peculiar mixture of thermal and nonthermal (see Chapter

1) radiation. The arched filaments and the core seem to be primarily thermal, while the filaments which cross the galactic plane (equator) are nonthermal in origin. The material in the clumpy spiral-like structure at the heart of Sgr A is mostly thermal, although the compact source, Sgr A*, is nonthermal, which is typical of compact sources in radio galaxies and quasars.

Infrared (IR) observations have provided an additional gold mine of data on the thermal emission from the galactic center. The data reveal 17 IR sources, most of which are apparently associated with the nucleus and whose combined emission is equivalent to about 10 million suns. Several of the IR sources are identified as being old giant stars, whereas another, known as IRS 16, may be a cluster of older stars very closely associated with the dynamic center of our galaxy, although its position does not quite coincide with Sgr A*. IRS 16 may be locked in an enduring embrace around Sgr A*. However, if IRS 16 is a cluster of stars and Sgr A* a black hole, the situation could be reversed, depending on the mass of the black hole.

IR observations of spectral lines produced by ionized neon reveal a lot of motion (up to 700 km/s) in the hot gas near the nucleus, suggesting that very energetic events (explosions perhaps) have triggered the chaotic movement.

X-ray emission from the galactic nucleus is centered on Sgr A. About 12 other X-ray sources (as yet unidentified, but possibly clusters of stars known as T Tauri stars) are located in the neighborhood. A very intense spectral line in the gamma-ray region of the spectrum had been observed since 1970, but it switched off in 1980. This source is regarded as strong evidence for a possible black hole at the galactic center, although the mechanism that shut off the source remains a mystery.

## Atomic and Molecular Clouds

Radio waves are emitted by a host of molecules (combinations of atoms) as well as several atoms, such as hydrogen (see Chapters 9–11). In the galactic-center region the molecule carbon monoxide (CO) is an important tracer of activity, but in the subsequent discussion the molecular and atomic data are not specifically singled out for attention; suffice it to say that these gases are seen in the nucleus and observation of their motion helps formulate the picture presented.

Infrared observations of the spectral signatures from several atoms, such as neon, are also a prime source of data upon which information on motions in the galactic nucleus is based.

## The Very Center and the Black Hole

The scale of the radio filaments protruding from Sgr A (Figure 6.2) is about 3 light-years wide by 130 light-years long. The parallel filaments lie about 200 light-years from the center, and their relationship to the nucleus is not obvious.

The filaments may be a huge arc of magnetic fields along which electrons are streaming in much the same way that a current flows along a wire, although, if this is true, the source of the electrons is quite unknown.

The central region can best be described in the following way. Within 10 to 15 light-years a very distinct ring of gas, containing dust and molecules, is seen nearly edge-on and tilted with respect to the galactic plane. Its average radius is about 10 light-years and it is rotating at 110 km/s. At the inner edge of this ring (radius 2–3 light years) the gas is very hot and ionized by the stars close to the center, and is a strong emitter of IR radiation. Inside this ring is a relatively empty region of space, devoid of gas and dust, but with increasing numbers of stars toward Sgr A* which may, in turn, be driving gas outward and creating the hollow region. The gas in the ring is clumpy and is at 5000 degrees K. A barlike segment (at a temperature of 12,000 K—inferred from observations of a variety of spectral lines) crosses the ring, and gas seems to be streaming out along the bar from some invisible source at the center. However, the Sgr A* radio source and the IRS 16 infrared object are located at one end of this bar, on the edge of the ring, and cannot be the cause of this outflow.

The source of ionization along the inside edge of the ring is inferred to be at a temperature of 35,000 degrees K. Such high temperatures can only be produced by very hot stars. These stars, in turn, burn their fuel very rapidly, and hence they are also very young objects. One interpretation of the data suggests that a burst of star formation may have occurred 10 million years ago and that most of the very massive stars have long since exploded and the ejected matter may now be falling back in, producing the spiral-shaped region of the gas ring. Disruption of the clouds close to the center would explain why the IR observations show only older stars. Recent star formation may have been inhibited.

The gas ring cannot last long, since the motions observed within it are so great that the ring should disrupt in 10,000 years or less. This supports the notion that the gas clouds will not be around for long because they are spiraling in to be swallowed up by a black hole. A black hole at the galactic center would account for the gamma-ray emission, as well as the intensity of the Sgr A* radio source.

New radio observations with very high resolution and studies of infrared spectral lines from the galactic center appear to have revealed very strong additional evidence for a black hole. The motion of the spiral arc of radio emission to the right of Sgr A*, shown in Figure 6.3, appears to be part of a ring rotating at 110 km/s. The radio arc on the other side appears to be rotating at 137 km/s around Sgr A*. Whenever an object or a structure such as this ring is in rotation, it is possible to calculate the amount of mass at the center of the object, because this mass determines how fast the material will be rotating about it. For the galactic-center region an object of four million solar masses within five light-years of the center is required. Other IR observations of the nucleus show that the IR radiation is far more concentrated than would be expected if it were due to the combined radiation of even thousands of stars in a star cluster. Hence the central object is almost certainly a black hole of four million solar masses.

Sgr A* appears to be surrounded by an ionized region of relatively low density in which a stellar cluster of two million solar masses may exist. In turn, the ring of dense gas, its inside edge mostly ionized, rotates. This disk is clumpy and extends from 5 to 30 light-years out. Motion within it suggests that its structure has been severely disrupted during the last 100,000 years or so. Depending on the availability of gas, infall of material is expected toward this central black hole. At this time, though, the region appears to be relatively quiet, making the galactic center very subdued as compared to flare-ups that must have occurred in the past.

The detailed maps of the galactic center indicate the presence of a wide variety of phenomena, and when radio telescopes become even more adept at delineating fine-scale structures, our vision of the galactic nucleus will become even clearer. Jansky's detection of radio emission from Sagittarius and Reber's mapping of the radio sky began a remarkable journey of exploration into the galactic center. Their pioneering work also revealed that the entire Milky Way band of stars was the source of an extensive radio glow.

# 7

# The Milky Way Radio Beacon

## "A Steady Hiss Type Static of Unknown Origin"

In the summer of 1931 Karl Guthe Jansky began a study of "atmospherics" using his "merry-go-round" antenna (shown in Figure 16.1). The "atmospherics" were unexplained noises which interfered with newly installed transatlantic radiotelephone circuits; Bell Telephone Laboratories, Jansky's employer, wanted to know the source of the sounds. The "merry-go-round" was a rotating antenna which allowed some directionality to be obtained in the search for the unwanted static. Jansky's antenna and associated receivers operated at a wavelength of 30 meters (frequency of 10 MHz).

In the first report of his experiments Jansky wrote that he detected three separate groups of static; from local thunderstorms, from distant thunderstorms, and "a steady hiss type static of unknown origin." Little did he know as he planned to pursue the search for the origin of this static that he was fathering a new science.

In his second report, published in 1933, Jansky concluded that the source of the steady hiss must be somewhere outside the earth since it seemed to move through the sky along with the stars in a manner consistent with its being of an extraterrestrial nature. He established an approximate direction for the source as 18h right ascension and $-10°$ declination.

In a third report, published in 1935, Jansky concluded that the

. . . radiations are received any time the antenna system is directed toward some part of the Milky Way system, the greatest response being obtained when the antenna points toward the center of the system.[1]

Within Jansky's experimental inaccuracies he found the peak radio emission to be located more or less in the constellation of Sagittarius. He attempted an explanation for the mechanism that generated the radio signals, suggesting that stars or interstellar matter might be the cause. We now know that cosmic-ray

---

[1] K. G. Jansky, *Proceedings of the Institute of Radio Engineers*, Vol. 23, p. 1920 (1935).

electrons spiraling about interstellar magnetic field lines produce the bulk of the so-called *radio continuum* from the Milky Way. Jansky also noted that the hissing sound of the radio waves from space was very similar to the hiss produced in his headset connected to the radio receiver.

## Receiver Noise—Listening to Radio Sources

The reference to "radio" in the term "radio astronomy" sometimes triggers visions of radio astronomers sitting beside a loudspeaker listening to cosmic music. However, it is only fruitful to listen to the sounds emerging from the receivers connected to the radio telescope when trying to identify a source of unwanted radio interference. Radio astronomers never actually listen to sounds generated by their radio receivers.

Cosmic radio signals exhibit a characteristic hiss like that in a home radio when it is tuned between stations, or in a television when it is set to an unused channel. This hiss is called *noise* and is electrical in origin, being produced by random movement of electrons inside the electronic components of the radio set. The noise generated within a television set can also be seen, as "snow" on the screen.

Radio waves from cosmic sources are generated by the motion of electrons, either traveling close to the speed of light (relativistically) or more slowly (nonrelativistically). In either case the random movement of the electrons in radio sources creates electrical (or more correctly, electromagnetic) noise which is indistinguishable from noise produced by the receivers attached to the radio telescope. Just as it is difficult to hear someone speaking above the noise of a crowded cocktail party, the presence of internally generated receiver noise makes the detection of the radio whispers from space very difficult.

One of the greatest challenges electronic engineers confront in building radio astronomy equipment is to reduce the noise generated within the electronic components. The task of construction of low-noise receivers has now been elevated to an art, with a considerable fraction of the budget for new radio telescopes set aside for the development of highly specialized low-noise receivers.

The experience of listening to random noise can be extremely soothing to the ear, as in the case of the sounds of distant surf or a waterfall, but listening to the noise from space is of little practical value. The cosmic radio signals need to be translated into electrical currents which are then used to drive a moving pen over a paper chart or converted into numbers to be handled by a computer for later study.

Radio astronomers, therefore, neither look directly at nor listen to radio sources. Instead, the radio signals from space are processed in computers and displayed in a way which means something to the human eye. At the same time, quantitative information concerning the intensity of the radio signal has to be derived through accurate calibration measurements and calculation of antenna and receiver characteristics. The training of many radio astronomers includes learning how to perform these functions and then interpreting what the data signify about events in space.

## Grote Reber Maps the Milky Way

Jansky could hear the faint radio hiss from space in his earphones and went further to report on his quantitative measurements of the intensity of the received emissions. However, his discoveries went largely unrecognized by astronomers, either because they never got to read Jansky's technical papers, which were published in a journal aimed at radio engineers, or because the astronomers, not familiar with radio engineering, simply were not interested. But a few people did take note and it was Grote Reber, resident of the Chicago suburb of Wheaton, Illinois, a radio engineer by profession and a radio amateur (ham), who built the world's first dish-shaped radio telescope (Chapter 16).

In the spring of 1938 Reber set his equipment to receive at a wavelength of 9 cm (frequency of 3300 MHz), but he had no success in finding the cosmic static. The wavelength of 9 cm was chosen because he expected the radiation from the Milky Way to be thermal in origin and therefore the sky should have been brighter at wavelengths shorter than that used by Jansky (30 meters). However, the intensity of the Milky Way at 9-cm wavelength was well below Reber's sensitivity limits, because the Milky Way does not emit thermal radiation. It was only realized nearly 20 years later that the actual radiation mechanism is nonthermal, and therefore the radio signals are weaker at shorter wavelengths.

Reber decided to try again at a longer wavelength and built a new receiver and antenna to operate at a wavelength of 33 cm. Again he did not detect any signals. Finally, in his third attempt, at a wavelength of approximately 2.4 meters (frequency of 125 MHz), he detected a signal from the Milky Way and began a systematic mapping of this "cosmic static." (His 31-foot-diameter paraboloidal dish, which he constructed single-handedly, is shown in Figure 16.2 in the later chapter on historical aspects of radio astronomy.) The resolution of his observations was 12.5 degrees, which may be compared with resolutions of one second of arc now commonly achieved.

Reber discovered that greater amounts of radio emission seemed to originate from specific directions, notably Cassiopeia, Cygnus, and Sagittarius, this discovery being the first hint that individual radio sources might exist.

## A Modern Radio Map of the Sky

In 1982, after 15 years of systematic observations with some of the world's largest single-dish radio telescopes in England, Germany, and Australia, a detailed all-sky radiograph (shown in Figure 7.1) was published. The image shows what the radio sky looks like at 408 MHz (73.5 cm) with a resolution of one degree. Three billion numbers and a half-million picture elements were

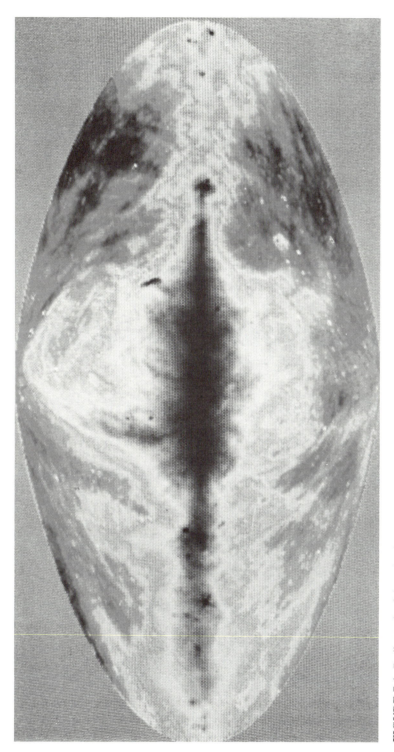

**FIGURE 7.1.** Radiograph of the galactic radio emission at 75-cm wavelength. This full-sky image is directly comparable to the optical photograph of Figure 6.1. The original radiograph was made in color; the dark region along the galactic plane represents the brightest radiation and the dark areas well removed from the Milky Way, upper and lower right, are regions of lowest emission. Notice the North Galactic Spur projecting to the north of the Milky Way just to the left of center. This map is the result of 15 years of work by a team led by Glyn Haslam using radio telescopes all over the world. (C. G. T. Haslam.)

required to produce this radiograph of the "steady hiss type static" from the Milky Way. The image represents one of the great achievements in modern radio astronomy. Figure 7.1 is presented in galactic coordinates and can be directly compared with the all-sky photograph shown in Figure 6.1. The strong emission band associated with the Milky Way is very evident. In the plane of the Milky Way a greater depth of the galaxy is intercepted by the radio telescopes; hence the intensity of the received radio waves from these directions is greater.

Figure 7.2 is a contour-map version of Figure 7.1. Here isolated radio sources are clearly visible as sets of concentric ring contours. A great number of radio sources lie in the galactic plane, and optical and related studies have identified them as either nonthermal galactic sources associated with the death of stars or nebulae associated with regions of star birth (see Chapter 8).

Several *spurs* of continuum emission are strikingly visible in Figures 7.1 and 7.2. The North Polar Spur, projecting from the Milky Way at longitude $30°$, is believed to be a segment of a huge radio-emitting shell ejected a few million years ago by a star that exploded within a few hundred light-years of the sun. This spur can be followed to latitude $+80°$ near the north galactic pole. A much smaller spur exists just above the galactic center (coordinates $l = 0°$, $b = 0°$) and is lined up with the protuberance described in the previous chapter (Figure 6.5), reaching to galactic latitude $+20°$.

The region of complex structure around $l = 80°$, $b = 0°$ in Figure 7.2 is known as Cygnus X. Here a mixture of thermal and nonthermal radio sources is observed along a galactic spiral arm viewed end-on in the direction of the constellation Cygnus (the Swan). Another spiral arm is seen edge-on in the constellation of Vela, around $l = 265°$, $b = 0°$.

The radio galaxy Cygnus A is clearly visible at $l = 76°$, $b = +6°$ while Centaurus A, the radio galaxy discussed in Chapter 3, is visible at $l = 310°$, $b = +20°$. The bright contours at $l = 280°$, $b = -32°$ are produced by the Large Magellanic Cloud, a nearby galaxy 150,000 light-years away. The patch of emission around $l = 208°$, $b = -18°$ is associated with the Orion nebula, the central faint "star" visible in the sword of Orion. The strongest radio source in the sky, located at $l = 112°$, $b = -2°$, is associated with the remains of a star that exploded three hundred years ago in the direction of the constellation of Cassiopeia. Its portrait will be shown in Chapter 8.

## Polarization of the Galactic Radio Waves

The hypothesis that the broad band of emission from the Milky Way is produced by cosmic rays spiraling around large-scale interstellar magnetic fields is supported by observations of the spectrum and the polarization of the emission. Large-scale surveys of the polarization of this radiation have been made at three wavelengths to allow Faraday rotation effects (see Chapter 3) to be removed. Figure 7.3 is a rather complicated-looking diagram which represents the results of nearly two decades of work to measure this polarization. Each small line

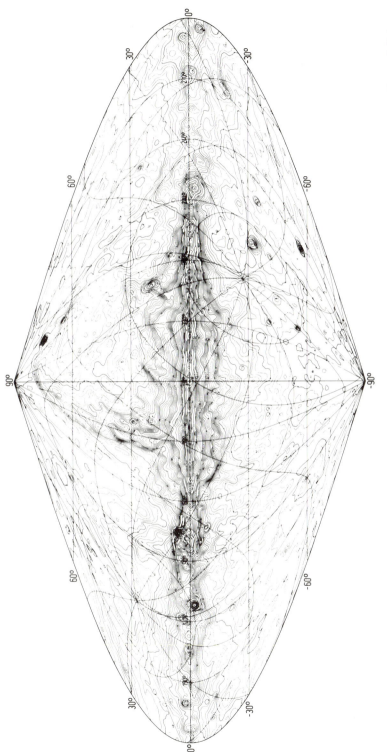

**FIGURE 7.2.** All-sky radio contour map of the data shown in Figure 7.1, at 408 MHz. The locations of bright radio sources (concentric contours) as well as the North Polar Spur, the band of enhanced emission north of longitude 30°, are clearly evident. Centaurus A is visible on this map at longitude ($l$) = 312°, latitude ($b$) = + 20°. Cygnus A (Figure 3.5) is at $l$ = 76°, $b$ = + 6° and Cassiopeia A (Figure 8.6) is at $l$ = 112°, $b$ = − 2°. (C. G. T. Haslam, C. J. Salter, H. Stoffel, and W. E. Wilson.)

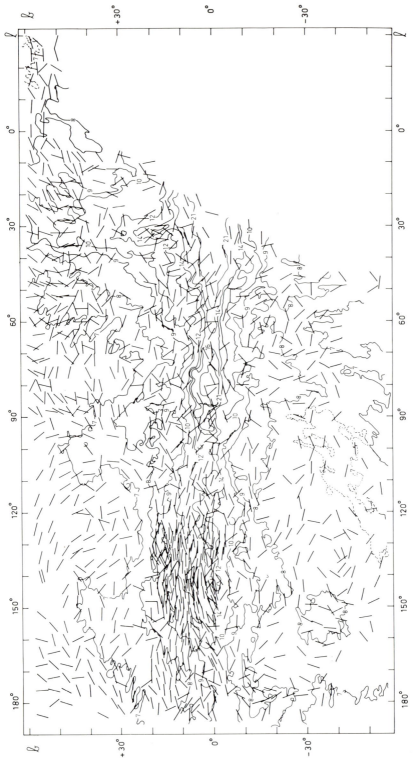

**FIGURE 7.3.** A map of part of the sky, indicating the interstellar magnetic field direction derived from radio polarization observations. The short, straight lines indicate the magnetic field direction at each location. The field appears most highly organized at longitude 140° in the plane of the Milky Way itself, latitude 0°. The polarization information is overlaid on the 75-cm-wavelength contours, similar to those given in Figure 7.2. (T. A. Th. Spoelstra, Dwingelo Observatory.)

on the map represents the direction of the magnetic field in the radio-emitting region, typically located within a few hundred light-years of the sun. A high degree of order in the magnetic field is seen around longitude 140°. In other parts of the sky the field is less ordered.

## The Appearance of the Radio Sky

Because of the nature of the nonthermal spectrum, the radio sky appears brighter as the observing wavelength is increased. At long wavelengths the entire sky appears to glow, and as the observing wavelength is increased the Milky Way band becomes broader and broader. The trend continues until the Milky Way is no longer brighter than the rest of the sky. Instead, at wavelengths of many tens of meters, first the Milky Way and then the entire heavens grow darker where patches of nearby thermal gas between the stars begin to absorb the "background" nonthermal radio signals. The appearance of the radio heavens, therefore, differs quite a bit at different wavelengths.

At no wavelength does the radio sky look like the optical sky. None of the few thousand stars we can see at night are radio emitters of any significance. Except for the Orion nebula, the Magellanic Clouds, and the Andromeda galaxy, none of the radio sources seen in Figure 7.1 or 7.2 are visible to the naked eye.

## "Normal" Galaxies

Our galaxy is believed to be relatively normal, just like billions of others in our universe. We might therefore expect most other galaxies to emit weak radio signals, and that is precisely what is found. However, most of them are so far away, and therefore appear so faint, that they do not show up even in the most sensitive studies of the radio sky. By comparison with radio galaxies and quasars, normal galaxies are all but invisible to radio astronomers.

The nearby spiral galaxies such as M31 in Andromeda or M33 in Triangulum, members of the Local Group, are exceptions because they are so close, only a few million light-years away. A radio map of M31 will be discussed in Chapter 9.

Figure 7.4 shows a radiograph of a very distant, apparently normal spiral galaxy, NGC 2276. The radiograph shows the spiral pattern of this galaxy to have been pushed inward at one edge, where optical studies have revealed the presence of lots of young stars. This phenomenon may be related to collision with an intergalactic medium through which the galaxy is moving, the interaction apparently having triggered a burst of star formation.

Another pathological case of a "normal" galaxy, this one NGC 4258, is shown in Figure 7.5. Here the radio "emitting arms" do not even overlap the optical features. This remarkable state of affairs may be the result of gas ejected

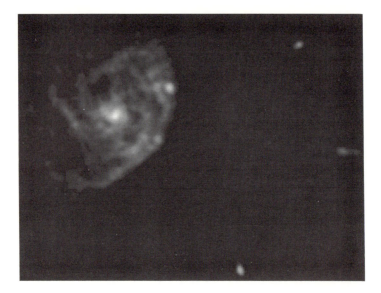

**FIGURE 7.4.** Radiograph of the spiral galaxy NGC 2276, showing faint radio emission from the spiral arms. The highly asymmetrical structure of this galaxy is believed to result from the galaxy's ploughing through intergalactic material, pushing up against the galactic gas. In optical photographs large numbers of very luminous emission nebulae are visible where the interaction with intergalactic material is believed to be taking place. Several faint background radio sources are also seen in this 20-cm-wavelength radio image; resolution, 4 arcseconds. Vertical size is 4'. (NRAO. Observer—J. J. Condon.)

from the nucleus interacting with matter, a "halo" of material around the galaxy. Radio emission from the "normal" spiral structure is apparently too faint to be detected on earth. These examples help to illustrate the thesis that upon closer examination there may be no such thing as a "normal" galaxy!

Jansky's observations of the cosmic static led to the birth of radio astronomy and the subsequent discovery of such remarkable objects as radio galaxies and quasars and the Sagittarius A radio source at the galactic center. All-sky surveys have drawn attention to a remarkable variety of peculiarly interesting radio-emitting objects within our galaxy, each a character in its own right, each with its own distinguishing shapes and features, many revealing the existence of quite mysterious phenomena occurring in the invisible universe around us.

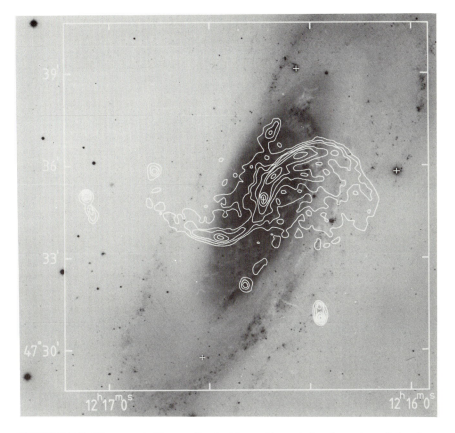

**FIGURE 7.5.** The remarkably peculiar-looking radio emission from the spiral galaxy NGC 4258 superimposed on an optical image of the galaxy (photo by A. Sandage). The radio structure appears unrelated to the ''normal,'' optically visible structure. The anomalous spiral ''arm'' is probably produced by an interaction between gas ejected from the nucleus with gas in the disk of the galaxy or gas in a more extended ''halo.'' (G. D. van Albada, J. M. van der Hulst, and J. H. Oort, Leiden Observatory.)

# 8

# The Galactic Radio Nebulae

## The Supernova—Star Death

On a chi-chou day in the fifth month of the first year of the Chih-ho reign period a guest star appeared in the south east of Thien-Kuan measuring several inches. After more than a year it faded.[1]

With these words Chinese astronomers in the year 1054 A.D. recorded their observation of a new star in the constellation Taurus. Early in the morning of that same day, July 5, Plains Indians in the western U.S. may have witnessed the appearance of this new star, an event so startling to them that they etched a record of it into the rocks. Figure 8.1 shows a pictograph found in Chaco Canyon, New Mexico. A starlike symbol, seldom used by these Indians, is located close to a crescent. The new star described by the Chinese astronomers, who had a long tradition of observing the heavens, did indeed appear close to the crescent moon, and it would have been visible in the early morning from the rocky overhang in Chaco Canyon.

Today we call such a guest star a *supernova,* and recognize the phenomenon as the violent destruction of a star in its final minutes. The remains of the supernova of 1054, which was visible during the day for several months, are now visible as the *supernova remnant* known as the Crab nebula, or Taurus A, one of the brightest radio sources in the sky. The radiograph of Taurus A is shown in Figure 8.2a. The radio portrait shows a display of luminous filaments very similar to those in the optical image (Figure 8.2b), showing that these filaments emit both radio and light energy. (Note that the appellation "nebula" refers to cloudlike objects in space. They all look like fuzzy clouds when viewed through small telescopes.)

The Taurus A radio source was already associated with the Crab nebula in 1947 because it was so obvious in photographs. This object became a primary testing ground for the theory of synchrotron emission. The light and radio emission from the Crab is polarized, and the spectrum of its radiation is clearly

---

[1] Ho Peng Yoke, *Vistas in Astronomy,* Pergamon, London, 1962, p. 127.

**FIGURE 8.1.** A rock painting (pictograph) in Chaco Canyon National Monument, possibly depicting the supernova of 1054 A.D. which is known to have occurred close to the waning moon and would have been visible from Chaco Canyon just before sunrise on July 5. The supernova gave rise to the Taurus A radio source shown in Figure 8.2. This painting is on the underside of a small overhang where it has remained protected for hundreds of years. (Von Del Chamberlain.)

nonthermal. Careful mapping by optical and radio astronomers showed that the magnetic fields in the Crab are aligned along the optically visible filaments. Today, more than 900 years after the explosion, the Crab continues to radiate. And the reason it does so has to do with a very odd star that spins furiously at the center of the nebula. This object, known as a pulsar, will be described in Chapter 11.

## Recent "Guest Stars"

In late August of 1975 a new star, almost as bright as Polaris the Pole Star, appeared just beyond the tail of Cygnus the Swan. Eight days later it faded from sight, and by then it was officially called Nova Cygni 1975. It wasn't a supernova, but something less dramatic—a nova. The word nova means "new star." Unlike a supernova, which is the complete disintegration of a star at the end of its life, a nova is a relatively gentle explosion which tosses a cloud of gas away from the star's surface in a small upheaval which does not shatter the star. Some novae are recurrent, repeatedly convulsing and ejecting hot

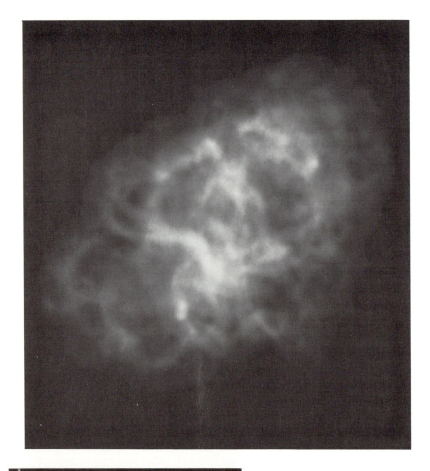

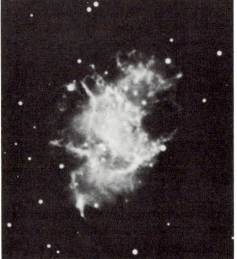

**FIGURE 8.2.** Radiograph (top) and optical photograph (left) of the Crab nebula (Taurus A) supernova remnant observed at a wavelength of 20 cm. The object is 6000 light-years distant and about 6 light-years across. A pulsar is located at the center of this remnant and continues to supply it with energy so that now, 900 years after the explosion, the nebula still shines. Vertical size (top) is 4′. (NRAO. Observers—A. S. Wilson, D. E. Hogg, and N. Samarasinha.)

clouds of gas which flare and cause the star to become a million times brighter for a week or so. Then the nova fades almost to its normal brightness over the next few decades. A supernova, however, shines with the light of a billion suns, and then, over several months, slowly fades away, although its remnant may still be visible to large telescopes for thousands or tens of thousands of years afterwards.

In 1811 A.D. Chinese astronomers observed a "guest star," visible for six months. It was a genuine supernova and its remnant is now a radio source, 3C 58, whose portrait is shown in Figure 8.3. More recently, in 1572, Tycho Brahe, the famous Danish astronomer, recorded a supernova which was seen all over the world. Its remarkable radio image, shown in Figure 8.4, reveals a beautiful ring structure produced by a shell of matter exploding outward.

Not since 1604 A.D. has a supernova in our galaxy been observed. In that year Johannes Kepler, the astronomer famous for discovering the laws which govern the movement of the planets about the sun, recorded the appearance of a supernova that was clearly visible to the naked eye. Despite nearly 400 years of astronomical research and development of sophisticated telescopes since then, no star within our galaxy is known to have exploded within eyesight. This fact is a source of frustration to astronomers who would dearly love to monitor

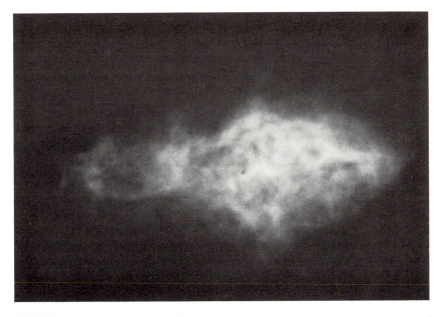

**FIGURE 8.3.** Radiograph of 3C 58, the remnant associated with a supernova observed in 1181 A.D. by Chinese astronomers. The distance to this object is uncertain and may be either 25,000 or 75,000 light-years. It appears to resemble the Crab nebula in that the remnant may have a filamentary structure. Vertical size 7′. (NRAO. Observers— S. P. Reynolds and H. D. Aller.)

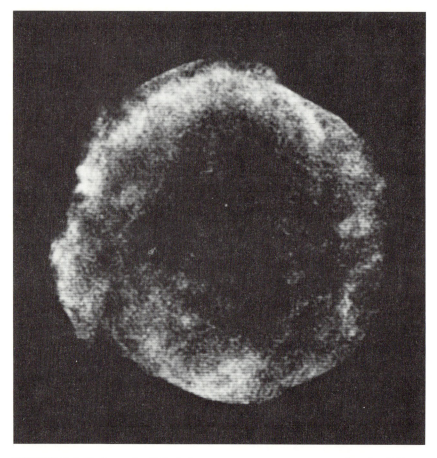

**FIGURE 8.4.** Radiograph of Tycho's supernova remnant at a wavelength of 10 cm, made with the Cambridge University 5-km radio telescope; resolution, 4 arcseconds. Vertical size is 10′. (D. Green and S. Gull, Mullard Radio Astronomy Observatory.)

a star in its last years, especially a star that is relatively close to us and which might be studied for some time before it exploded.

One supernova per 50 years in a typical galaxy is estimated to be a reasonable average rate for such events, based on observations of hundreds of supernovae in other galaxies. The Milky Way appears to be long overdue for the next one. Someday, perhaps soon, the world will be stirred by the appearance of another "guest star," and when it flares in our skies it will draw more attention than any previous astronomical event, including the visit of Halley's comet. No one can say when or where a hapless star will disintegrate and reveal the secrets of star death to the astronomer's relentless gaze. However, until we observe a good before-and-after case we cannot be certain that the theories (based on very good computer simulations of stellar birth, life, and death) of supernovae are correct.

All the supernovae described above are located many thousands of light-years from us and all of them are in the Milky Way. Perhaps, even today, somewhere in our galaxy, a star is shattering in its final agonies, but we may not know about that until the light has journeyed the thousands of light-years that separate us.

# Cassiopeia A

The brightest radio source in our heavens, other than the sun (which appears bright because it is so close to us), is a supernova remnant in the constellation Cassiopeia. First efforts to identify an optical object at its location were rather disappointing. Nothing as blazingly obvious as the Crab nebula was found. Only very faint filamentary material could be seen on photographs and it appeared that this was indeed the remains of an exploded star. The presence of dust between the sun and the remnant 10,000 light-years away may be preventing us from seeing the rest of the remnant, but radio waves ignore such dust and have revealed this object to be a glorious display of radio-emitting filaments.

In the 1960s radio astronomers mapped the Cassiopeia A radio source with the first radio telescope systems that combined several dishes to simulate a single, larger dish. Figure 8.5 is an early contour map of Cas A (as its name is abbreviated) superimposed on a negative print of an optical photograph of the same region of sky. The optical nebulosity can be seen as small black patches in this negative image. The radio emission from Cas A appears to be ringlike, suggesting a shell of material ejected by an explosion. Observations of the motion in the filaments indicate that the explosion must have occurred in 1667 A.D., but no record exists of anyone having seen this event.

Over the last several years the Very Large Array (VLA) as well as the Cambridge University 5-km radio telescope have been used to make faithful portraits of Cas A. One such image is shown in Figure 8.6, which reveals a magnificent display of cosmic fireworks, the Cas A remnant 300 years after the explosion. This image is surely one of the most dramatic radiographs ever made.

The very observant reader might notice what appear to be a series of very faint concentric circles in the radiograph. This is due to a tiny error associated with one of the VLA's 27 radio antennas involved in making the observations. After the error is corrected a "perfect" image will be produced. However, the production of a radiograph such as this, with relatively limited computing power, is an enormous task. Forty hours of observations with the VLA, involving 5 million independent measurements stored on 20 magnetic tapes, were required. For computer aficionados, it would take a VAX 11/750 (commonly used at radio observatories), working 24 hours a day, 48 days to produce the radiograph.

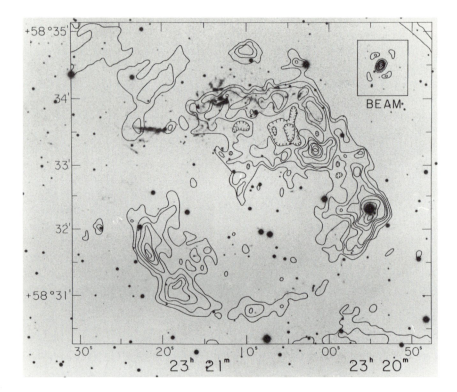

**FIGURE 8.5.** This contour map of the Cassiopeia A radio source, produced in the late 1960s, is overlaid on an optical (negative) photograph taken with the 200-inch telescope of the Mt. Wilson and Palomar Observatories by S. van den Bergh. A few faint optical filaments are seen. The radio data suggest a ringlike structure for Cas A. (NRAO. Observers—D. E. Hogg, G. H. MacDonald, R. G. Conway, and C. M. Wade.)

The making of the Cas A image required the use of a Cray supercomputer, not yet commonly available to radio astronomers. The National Radio Astronomy Observatory (NRAO) rented time on a Cray supercomputer in Hollywood in order to make this radiograph. That machine generates many of the special graphics effects for high-budget science fiction movies. Perhaps it was appropriate that radio astronomers worked in Hollywood to produce this spectacular image. The Cygnus A radiograph, seen in Figures 1.2 and 3.5, was also produced by the Cray.

## Supernovae of Type I and Type II

The theory of supernovae is reasonably well developed. Fortunately the sun does not appear likely to suddenly explode violently and wipe out all terrestrial life. On the other hand, it now seems fairly certain that when the sun does

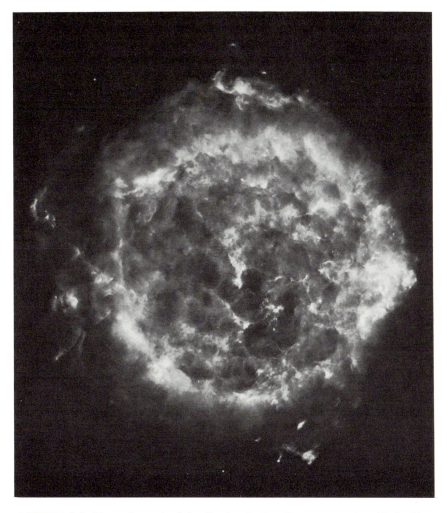

**FIGURE 8.6.** The radiograph of the Cassiopeia A radio source made with the Very Large Array radio telescope at 20-cm wavelength. Resolution, 1 arcsecond. Comparison with Figure 8.5 highlights the dramatic progress that has been made in radio astronomical image-making capability in the last 15 years. The bright, radio-emitting filaments are produced by magnetic fields illuminated by cosmic-ray electrons traveling close to the speed of light. Vertical size is 4′. (NRAO. Observers—R. J. Tuffs, R. A. Perley, M. T. Brown, and S. F. Gull.)

enter its final phases of life, about five billion years from now, the earth will be destroyed. The irony regarding supernova theories, however, is that they have not been tested by direct observation because no one has observed a star before and after the event and, therefore, no one can be absolutely sure which types of stars explode in this manner. (Although astronomers, especially the theoretically inclined, believe in their theories, direct comparison with the real universe often leads to surprises!) Astronomers have witnessed a regular display of supernovae in other galaxies, where the flash may briefly outshine the entire galaxy before fading from view, but the stars in those galaxies are too faint and far away to have allowed any of them to be recognized as individuals from earth.

Stars containing more than four solar masses of gas are likely to explode at the end of their lives, but not all of them will do so in the same way. The Crab nebula and Cassiopeia A represent a different class of event as compared to Tycho and Kepler's supernovae. The latter are examples of Type I supernovae, believed to be the destruction of what was originally a white dwarf star—a highly mature star which, in its dotage, shrinks to a mere shadow of its former self. These white drawfs were probably members of binary star systems. (Approximately half the stars in our galaxy are paired in binaries, unlike our sun which has no close companion.) The interaction between binary stars can be very dramatic. If one member is a dwarf star, evolving slowly, and the other member a more massive star, aging rapidly, the larger star may enter a phase in its life when it swells to enoumous size. Some of its material then falls onto the dwarf, and if this process continues for long enough, the white dwarf may suddenly be incapable of absorbing any more of its neighbor's refuse, overheat, and explode. The Tycho and Kepler supernovae are believed to have been produced in this manner. They produce beautiful, luminous shells of emission. Study of the light from these supernovae gives much information about the chemical constituents of the exploding material, and the data show distinct differences from the other types of supernova. Also, Type I supernovae are regularly observed to occur in distant elliptical galaxies which contain no interstellar matter, nor any young, massive stars capable of becoming supernovae on their own.

A Type II supernova, on the other hand, may involve the explosion of a single massive star due to a catastrophic increase in the amount of heat generated in its core as part of its ''normal'' evolution. The Crab nebula is an example of such an event, which gives rise to filled-center sources as opposed to the shell-like structures of Type I. Cassiopeia A may be a pathological Type II whose explosion may have been less bright, which would explain why it wasn't visible on earth.

## Other "Radiogenic" Nebulae

Australian radio astronomers, with relatively limited resources, have performed sterling radio astronomical studies of the southern skies (see Chapter 16) and for several years have been studying dozens of supernova remnants with radio telescopes modified from their original function of observing the sun. At Fleurs, a radio telescope using 64 small 19-foot dishes was combined with four slightly larger dishes to make radiographs which involved two eight-hour periods of observation per radio source. As is so often the case with modified equipment, special hi-tech solutions were required to assure that the system worked perfectly—in this case, a lot of running between the dishes to personally check that they were still tracking the radio source. This dedication to the exploration of the invisible universe produced dramatic radiographs and healthy radio astronomers. At Molonglo, a second high-resolution modified solar radio telescope, named MOST, operating at 36-cm wavelength, was used to produce the images in Figures 8.7 and 8.8. According to the astronomers involved, the tracking capability of this telescope was much better and allowed the work to be pursued in a more relaxed, MOSTLy asleep, mode.

Figure 8.7 shows the supernova remnant known as G326.3−1.8, intermediate between Types I and II discussed above. It shows a beautiful shell as well as a bright center of emission within it. The exposure in this print is such that the brighter parts of the central source are burnt out.

The supernova remnant G332.2+0.2, shown in Figure 8.8, appears, at first glance, to be a confused mess. It is actually one of the most remarkable objects yet observed. The bright overexposed blob at the right is a shell-like supernova remnant, as revealed in different exposures in the image-generation process, and this image highlights an amazing phenomenon. Some of the matter from the supernova explosion has blasted out of the shell in a thin jet and then expanded into a great plume, seen at the upper left half of the radiograph. The jet's velocity is about 10,000 km/s and the phenomenology is similar to a jet in a radio galaxy expanding to create an extended lobe of radio emission.

The rogues' gallery of supernova remnants produced by the Australian survey reveals each object as an individual: some ringlike Type I, some filled structures

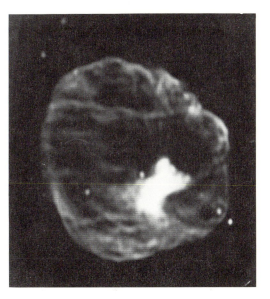

**FIGURE 8.7.** Radiograph of the supernova remnant G326.3–1.8, a composite of the two types of supernova remnants in one object. The shell-like Type I structure is dramatically evident and encloses a bright, filled structure which is supposed to be the hallmark of a Type II remnant. The bright feature is burnt out in this image. This object and that seen in Figure 8.8 were observed from Australia. Vertical size is 48′. (Division of Radiophysics, CSIRO. Observers—D. K. Milne, J. L. Caswell, R. F. Haynes, M. J. Kesteven, K. J. Wellington, R. S. Roger, and J. D. Bunton.)

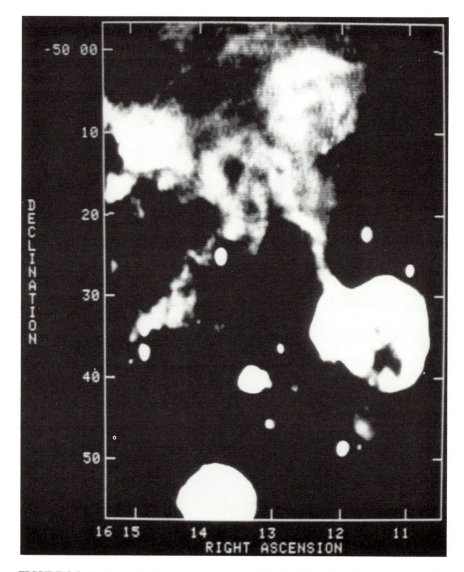

**FIGURE 8.8.** Radiograph of supernova remnant G332.2+0.2, which shows a jet emerging from the very bright blob at the right. In this exposure the details in the bright source are lost, but it is a shell-like remnant. The jet issues forth and then splays out to produce the broad, low-level plume in the upper left. (Division of Radiophysics, CSIRO. Observers— D. K. Milne, J. L. Caswell, R. F. Haynes, M. J. Kesteven, K. J. Wellington, R. S. Roger, and J. D. Bunton.)

of Type II, and several somewhere in between. Others, such as the object in Figure 8.8, do not fit any pattern.

Northern hemisphere observers have found their own quota of pathological objects. Figures 8.9 and 8.10 show the portraits of the Tornado nebula and the Bird nebula, two remarkable radio sources recently discovered in the general direction of the galactic center using the VLA. The discoverers at first wondered whether these were peculiar supernova remnants or a new class of radio source. If these are supernova remnants, their shapes have been dramatically distorted by interaction with surrounding interstellar gas. Both objects have bright, point-like sources at their boundaries, quite unlike normal (old-fashioned) supernova remnants. The Bird (Figure 8.10) shows a short, jetlike feature leading to the wings, which is similar to the patterns seen in Figure 8.8.

Figure 8.11 is a radiograph of another strange and as yet unidentified radio source suspected of being a supernova remnant. All these objects lie behind too much dust to allow any of their light to reach the earth.

## Supernovae and Life

The filaments within supernova remnants are illuminated by relativistic electrons spiraling around magnetic fields. As the remnant expands, ages, and runs out of energy, these electrons escape into surrounding space, where they become cosmic rays. These cosmic rays leak out into space, where they encounter magnetic fields between the stars. The electrons gain energy from the encounter

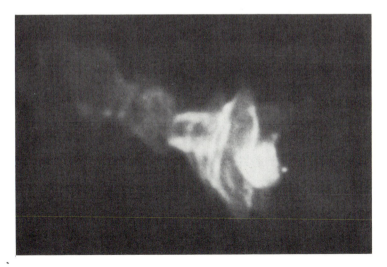

**FIGURE 8.9.** Radiograph of a new source, known as the Tornado nebula, discovered in the region of the galactic center. This is probably a supernova remnant, but is optically invisible. Vertical size is 12'. (R. H. Becker and D. J. Helfand. Reprinted by permission from *Nature*, Vol. 313, p. 115, 1985. Copyright © Macmillan Journals Limited.)

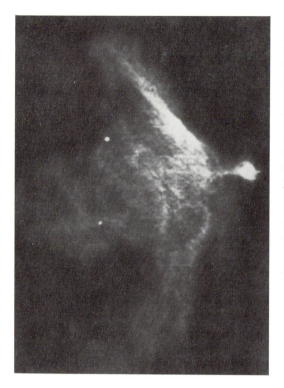

**FIGURE 8.10.** Radiograph of the Bird nebula, discovered in the region of the galactic center. The wings of the bird may be a plume-like region not unlike the feature shown in Figure 8.8. A very small jet appears to be present close to the source (bird's head) at the right. If this object is indeed a supernova remnant, it may be an unexpected variation on the radio jet theme, as may the object shown in Figure 8.8. Vertical size is 28'. (R. H. Becker and D. J. Helfand. Reprinted by permission from *Nature*, Vol. 313, p. 115, 1985. Copyright © Macmillan Journals Limited.)

and then, rejuvenated, transmit the synchrotron emission we pick up as radio waves from the Milky Way (Chapter 7).

It may take years for a remnant to fade from view. Figure 8.12 is a view of IC 443, over 10,000 years old and about 2000 light-years away. The extent of this nebula is about 15 light-years from side to side. The radio map was made at the low frequency of 151 MHz and shows that to one side the nebula appears to be piled up against an invisible barrier of interstellar gas while the other side is free to expand outward.

The total energy generated by supernova explosions over billions of years provides enough energy to propel interstellar clouds hither and thither, causing them to collide with each other. Supernovae keep interstellar matter stirred up, and when clouds collide new star birth is triggered. A nearby supernova remnant probably spawned the solar system.

Supernovae play a direct role in assuring our existence. The early universe contained none of the heavy elements, the basic constituents of matter which make up our world. Those elements include the oxygen and nitrogen we breathe and the carbon in every molecule of every cell in our bodies. Those elements, and thus all the atoms of which we are made (except the hydrogen), were formed in stars and were injected back into space and made available for future planetary formation because stars exploded. The heaviest elements were formed during the supernova explosions themselves. Every atom in a gold ring, for

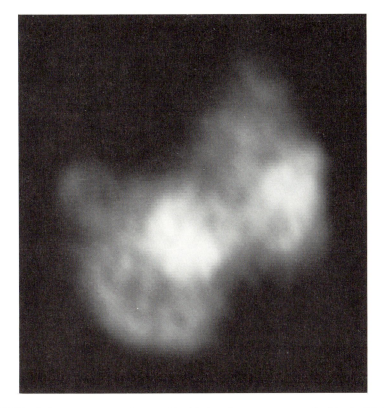

**FIGURE 8.11.** Radiograph of an unidentified radio source mapped with the VLA by Cambridge University radio astronomers. The source, known as G227.1 + 1.0, is probably a distant supernova remnant which is optically invisible because of the existence of opaque interstellar dust clouds between the object and the sun. Vertical size is 2'. (D. Green and S. Gull.)

example, was cooked up in the violence of star death and then traveled far, and for a very long time, before coming to temporal rest around someone's finger.

Today the material within the supernova remnant Cas A is being fed into space and sometime, somewhere, in the very distant future, some of those gases now radiating radio and light signals at us may be incorporated into the organism of some alien living entity.

## The Emission Nebula—Star Birth

Individual galactic radio sources were originally recognized as sets of closed contours (concentric circles) on maps such as that shown in Figure 7.2. The galactic radio sources, however, are not all supernova remnants, the aftermath of star death. Some of the galactic radio sources are associated with star birth.

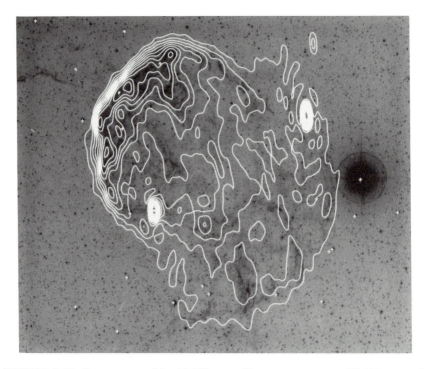

**FIGURE 8.12.** Contour map of the 10,000-year-old supernova remnant IC 443, mapped at 2-m wavelength with the new Cambridge University synthesis radio telescope, resolution 1.2 × 3.1 arcseconds, overlaid on an optical (negative) image of the nebula. Vertical size is 1°. (D. Green and S. Gull.)

In the constellation of Orion, visible in the mid-winter sky in the northern hemisphere, the famous hunter of the heavens presents three equally bright stars lined up to indicate his belt. The bright stars above and below the belt indicate his head, shoulders, and knees, but you can also see two or three stars outlining the sword hanging from Orion's belt. On a clear night you might be able to make out the middle star, and with binoculars you may see that it is not a star at all! It is a faint, fuzzy nebula. Located 1200 light-years away, this nebula glows because it has been heated to incandescence by ultraviolet radiation from four young stars within it. These stars were spawned from an interstellar gas-and-dust cloud about a million or so years ago, when an earlier, primitive ancestor of man roamed our planet. Those creatures would not have witnessed the birth of the Orion nebula as a sudden flaring in the sky such as accompanies a supernova explosion. On the contrary, the emergence from the interstellar womb is a gentle event. It is a process that occurs gradually, over a few hundred thousand years, perhaps. After the stars are born they may burn their way out of the dust and gas clouds which sheltered them throughout gestation, but even that will not cause the nebula to suddenly flare brightly in the heavens.

The Orion nebula has drawn as much attention from astronomers as any object in the universe and is the prototype of a class of radio source called an *emission nebula*. Such nebulae are incandescent, having been raised to temperatures of about 10,000 K by young stars within them, and they emit light, heat, and radio waves by the thermal emission process.

## HII Regions

Radio astronomers refer to emission nebulae as *HII regions*. The symbol "HII" refers to ionized hydrogen, while "HI" is used to represent cold or neutral hydrogen atoms, each of which consists of a single electron orbiting a proton. Hydrogen clouds exist everywhere in interstellar space (see Chapter 9). It is within such clouds that stars are spawned. When hot stars begin to shine, their highly energetic ultraviolet radiation streams outwards and shatters hydrogen atoms in surrounding space. The hydrogen atom is extremely sensitive to ultraviolet radiation, which kicks the electron out of its normal orbit about the proton and produces the ionized form, HII. Hot stars literally eat their way through surrounding gas cocoons within which they were born, converting what had been cold matter into hot gases which being to radiate their own energy. Within large emission nebulae many small, compact HII regions can form, each associated with a newly born star. Several compact HII regions are seen in the radiograph of W49 (Figure 8.13), a large HII region in which star formation is actively occurring.

The difference between a supernova remnant and an HII region is revealed by measurement of their spectra and polarization. The HII region has a thermal spectrum and its radio emission is unpolarized, while the radiation from supernovae is nonthermal and polarized.

Individual stars may range in size from giants (50 solar masses) to dwarfs (1% of a solar mass). The most massive stars burn their fuel most rapidly and are thus quickly exhausted. It is these stars which later explode as supernovae (within a few to 10 million years after birth). The more normal sunlike stars will shine for 10 billion years and will never be hot enough to create an HII region. Nor will such stars die violent deaths.

It is the massive stars which become supernovae of Type II, and since most massive stars are now known to be created in HII regions, we might expect that supernovae often occur inside HII regions. This is not observed! To put it mildly, this is one of the more interesting mysteries facing astronomers today.

## Planetary Nebulae

There are other galactic nebulae which are radio sources. Figures 8.14 and 8.15 are examples of planetary nebulae, shells of luminous matter ejected from a central star. These planetary nebulae, neither of which appears very shell-

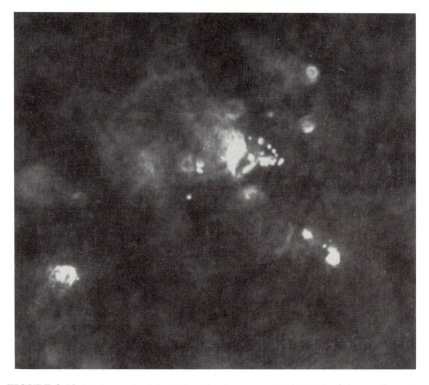

**FIGURE 8.13.** Radiograph of the HII region known as W49. Most of the small sources visible in this 6-cm-wavelength map are probably compact HII regions, some shell-like and others pointlike, immediately surrounding newly formed stars. This image is exposed so as to accentuate the structure; therefore, the low-level radiation is not as visible. The extent of the radiograph is about 5 × 4 arcminutes (cf. Figure 10.8). (J. W. Dreher, W. T. Welch and K. J. Johnston.)

like, are also located in the Milky Way. They represent far less violent affairs than supernovae, but the planetary nebula shares one common property with the former—they are both symptoms of star death. The supernova is the death of a massive star, while the planetary nebula signifies the death of a smaller, more normal star. The central star of the planetary nebula has merely shrugged off its outer layers. What is left of the inner parts of the star may collapse to become a white dwarf, and in due course it will fade from view and extinguish itself.

Sometime, about five billion years from now, the sun may shed its outer layers of hot gas and become a planetary nebula, swallowing the earth as it does so. The sun, just like everything in the universe, must die. Tens of thousands of years later, astronomers on alien planets may watch our planetary nebula

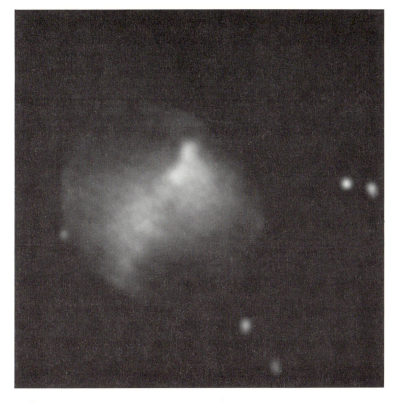

**FIGURE 8.14.** Radiograph of the planetary nebula NGC 6853, observed at 20-cm wavelength with the VLA. The radio image is very similar to the optical image of this object. (NRAO. Observer—R. C. Bignell.)

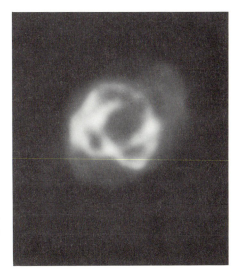

**FIGURE 8.15.** Radiograph of the planetary nebula NGC 6543, revealing two overlapping elliptical emission rings. (NRAO. Observer—R. C. Bignell.)

and study its radiation in order to better understand the nature of stellar evolution. Those beings will never know who, or what, orbited the original star.

Emission nebulae, supernovae, and planetary nebulae are constant reminders that we live in an extraordinarily dynamic universe in which everything changes and the cycles of birth, life, and death are all about us.

# 9

## Interstellar Hydrogen

### Clouds of Destiny

Once upon a time hydrogen gas was created in great profusion throughout the universe. Hydrogen is the basic building block of galaxies, stars, and nebulae. It is the element burned in the furnaces of the cores of stars everywhere. The conversion of hydrogen to helium in the so-called thermonuclear process keeps stars shining. It is in stellar interiors, where temperatures reach tens of millions of degrees, that hydrogen, and subsequently helium, is consumed and converted into heavier elements. These include carbon, oxygen, and nitrogen which, in their turn, become the building blocks of life.

Hydrogen makes up 80% of the universe's mass (the rest being mostly helium) and was formed soon after the big bang, the modern astronomer's version of creation (see Chapter 14). As the universe expanded and cooled, great volumes of this elemental gas formed huge clouds, millions of light-years across. These systematically shuddered into many smaller clouds, each only a few hundred thousand light-years in diameter, later to become the galaxies.

Within galaxies stars formed in great numbers as even smaller subunits broke free of the larger clouds and continued to collapse. Each collapsing cloud was pulled inward by gravity, which finally overcame the disruptive force of heat, that is, the internal energy of motion (known as kinetic energy). This process, a constant battle between gravity, which pulls a cloud in upon itself, and kinetic energy, which forces the cloud to expand, is witnessed everywhere in our universe. When gravity dominates, the cloud collapses to form a galaxy or a star, depending on the initial size of the cloud. But gravity can only overcome the disruptive influence of internal heat if the cloud cools enough by radiating away some of its energy as, for example, infrared radiation.

As the universe expanded and cooled, gravity became master of the destiny of gas clouds and determined the shape of things to come. Those shapes are what we now call galaxies, star clusters, stars, planets, and satellites such as our moon.

The space between the stars in the galaxy (interstellar space) is filled with hydrogen gas left over despite many generations of star formation. The hydrogen, in its basic neutral (or cold) form, known as HI, as opposed to HII which is the ionized form, drifts in great invisible swaths in space because neutral hydrogen gas does not radiate light.

During the Second World War a small group of astronomers in occupied Holland gathered to discuss topics of scientific interest. It was after this historic meeting that one of the scientists made the suggestion that because of the unique properties of the neutral hydrogen atom, it might be expected to transmit a detectable radio signal, which would mean that this important gas could at last be directly observed. Of course, no one yet had a radio telescope or sensitive receiver with which to make the search, but the scientists did know that pioneering radio astronomical work had been done in the USA before the war broke out. They would begin their search after the war ended.

## Generation of the 21-cm Spectral Line

The neutral hydrogen atom consists of a proton with an electron in orbit about it. Both the proton and the electron also *spin* about their individual axes, much as the earth spins on its axis. They can either spin in the same direction (called parallel spin) or in opposite directions (anti-parallel). The total energy contained by the atom in these two conditions is different. When the spin state flips from the parallel condition to the anti-parallel, which contains less energy, the atom gets rid of energy by transmitting a radio signal at one distinct wavelength close to 21 cm. This produces a *spectral line*. The 21-cm line is the characteristic signature of HI.

The uniqueness of this wavelength is a consequence of a basic law of physics which states that there is a direct relationship between energy and wavelength (or frequency). Known as Planck's law, it also states that any radio or light wave has a specific energy associated with it, depending on its wavelength. The shorter the wavelength, the higher the energy, which means that gamma rays are the most energetic form of electromagnetic radiation (see Figure A.3).

The term "spectral line" refers to the fact that if you adjust the tuning on a radio receiver operating around the hydrogen signal produced by a distant interstellar cloud (containing vast numbers of atoms), the radio message comes bursting through at just one wavelength, close to 21 cm (at a frequency of 1420 MHz). This is similar to a radio station on your AM or FM radio coming through at only one spot on the dial.

When the war ended, a scientific race began and in 1951 three research groups—in Australia, the U.S., and Holland—nearly simultaneously discovered radio emission from interstellar neutral hydrogen. This gas was found to be distributed in vast clouds between the stars. Even the solar system is surrounded by diffuse neutral hydrogen gas. Since then, the 21-cm line from all over the Milky Way, as well as from thousands of other galaxies, has been studied in great detail. The observations required elaborate data presentation techniques because the hydrogen clouds are all moving with respect to the sun, and thus the hydrogen line suffers a Doppler shift whose magnitude can be measured and which contains much valuable information about the nature of the interstellar gas.

## Observations of Interstellar Neutral Hydrogen

The HI clouds literally shine at a wavelength of 21 cm and this "cloudshine" is studied by radio astronomers. Within a hydrogen cloud individual atoms are moving with respect to each other, with the velocity determined by the temperature within the cloud. The spectral signal emitted by individual hydrogen atoms is therefore Doppler shifted to different wavelengths, depending on the motion of the atoms. As a consequence, the so-called spectral "line," which would otherwise be observed as a spike coming through on the dial, as it were, is spread over a small range of wavelengths (or frequencies). It is said to be *broadened* by random motion within the cloud. The width of the observed spectral line has long been believed to be a measure of the broadening, which is related to temperature in the cloud. This means that the radio telescope acts as a giant thermometer capable of probing into the well-being of distant parts of our galaxy. Cloud temperatures range from 10 K in small, dense clouds (densities hundreds of atoms per cubic centimeter of space) to 1000 degrees or more in large diffuse clouds (densities of less than one atom per cubic centimeter). These densities are all vastly lower than that of the air we breathe, which is about $10^{19}$ molecules per cubic centimeter.

If the electrons orbiting the proton in the hydrogen atom receive excessive transfusions of energy, for example, by being struck by ultraviolet light, they can be entirely separated from the protons. This produces ionized hydrogen (free protons) and may create an HII region such as discussed in the previous chapter. Alternatively, the electron may be kicked into a higher orbit, where it maintains contact with its parent proton. The atom is then said to be *excited*. If the electron later falls back into lower orbits it will radiate away energy in the form of a series of spectral lines known as *recombination lines*, with the wavelengths of these lines (mostly in the radio spectrum) depending on the energy differences associated with the changes in the electron orbit. The study of these recombination lines gives astronomers temperature and density information about the ionized hydrogen in HII regions.

## Radiographs of Interstellar Hydrogen

A dramatic breakthrough in the study of the large-scale distribution of interstellar hydrogen came in the mid-1970s, when surveys of the hydrogen in all directions in the sky were completed and modern computer graphics techniques were used to generate radiographs showing what the structure of interstellar HI looks like. Two such radiographs are shown in Figure 9.1. In these images the brightness

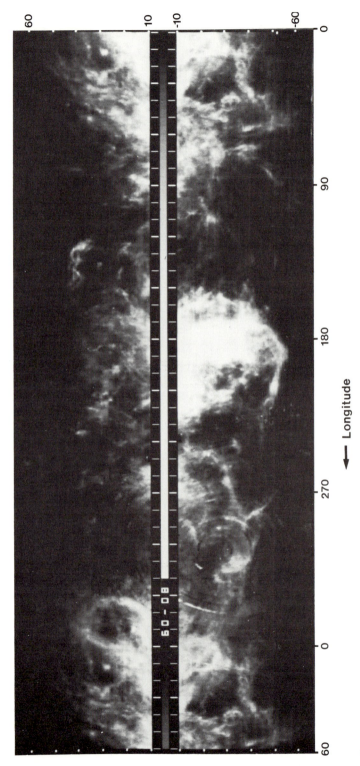

a

← Longitude

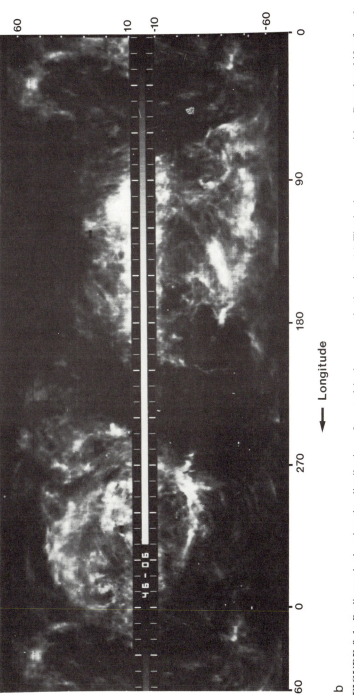

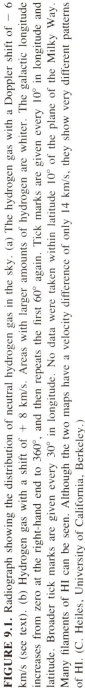

b

**FIGURE 9.1.** Radiograph showing the distribution of neutral hydrogen gas in the sky. (a) The hydrogen gas with a Doppler shift of − 6 km/s (see text). (b) Hydrogen gas with a shift of + 8 km/s. Areas with larger amounts of hydrogen are whiter. The galactic longitude increases from zero at the right-hand end to 360°, and then repeats the first 60° again. Tick marks are given every 10° in longitude and latitude. Broader tick marks are given every 30° in longitude. No data were taken within latitude 10° of the plane of the Milky Way. Many filaments of HI can be seen. Although the two maps have a velocity difference of only 14 km/s, they show very different patterns of HI. (C. Heiles, University of California, Berkeley.)

at any point is related to the total number of hydrogen atoms close to the indicated velocity in that direction. The greatest amounts of HI gas are found close to the plane of the Milky Way, but these maps show only the hydrogen above and below 10 degrees of galactic latitude, where there is relatively less gas and the structure is more clearly visible.

An exciting feature of these maps is the presence of enormous numbers of arcs and filaments (long streamers). These are visible as great threads of emission, whose shapes are almost certainly controlled by magnetic fields between the stars. Hydrogen gas pushed around by expanding supernova remnants which have long since faded into oblivion may be guided along magnetic field structures to produce the patterns seen in Figure 9.1. Recent research suggests that magnetic fields may play a far more important role in determining the structure of neutral hydrogen clouds.

## What Hydrogen Line Data Look Like

In the early days of radio astronomy the observations of many hundreds of rather complex 21-cm spectral lines were examined on an individual basis. When larger radio telescopes with far greater resolution and computers with very substantial data-handling capacity became available, the number of spectral lines the radio astronomer was capable of collecting became prodigious, and the data had to be processed in more sophisticated ways. Consider, for example, the Arecibo 1000-foot-diameter radio telescope, which has a resolution at 21-cm wavelength of 3 minutes of arc. To completely observe the hydrogen in the Milky Way over an area of, say, 5 degrees by 5 degrees would require that the spectral lines in 1000 directions be examined, an impossible task for one person. Instead, these spectra are used to make a contour map of the HI over the area of sky being surveyed. Such contour maps are plotted in a variety of ways. They might show the intensity of the HI emission at a given velocity, as does Figure 9.1, or they can display the HI emission as a function of position and velocity, where the telescope is essentially moving while the data are being collected. Figure 9.2 is an example of such a contour map which allows you to see hydrogen in depth, since velocity is an indicator of distance (see below). The position coordinate is produced by moving the telescope from point to point, in galactic latitude, for example, while keeping the longitude fixed. In each direction that the telescope is pointed a spectrum is recorded and these are then all used to make the final map. Such maps contain a very large amount of information and astronomers working with data in this form become used to thinking in a multidimensional space where one of the dimensions is velocity rather than position. The invisible universe of hydrogen gas then becomes "visible" to the imagination.

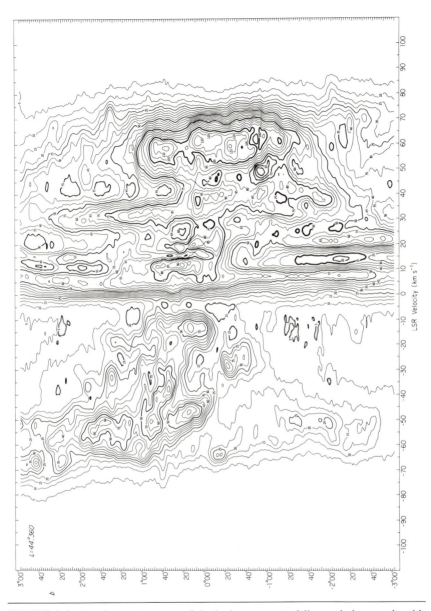

**FIGURE 9.2.** A radio contour map of the hydrogen spectral line emission, made with the Arecibo radio telescope (3 minutes of arc resolution) along a line which crosses the galactic plane at longitude 44.°36. The latitude range of the map covers 3° on either side of the galactic plane. The Doppler-shifted velocity is shown along the horizontal axis and is a measure of the distance to the emitting hydrogen gas. Contours represent regions of equal brightness of HI radio emission. The ridge running up the middle of the map is due to hydrogen close to the sun. Peaks due to emission from HI clouds and valleys, due to absorption by colder clouds, can be seen throughout the map. (T. M. Bania and J. Lockman.)

## Seeing into the Depths of Space

Motion within interstellar clouds means that the Doppler spread broadens the line; but, in addition, motion of the entire cloud, either toward or away from the sun, also produces a Doppler shift. For hydrogen clouds moving away from us the radiation is observed at a wavelength slightly longer than expected (redshifted) and for those coming toward us it is shifted to shorter wavelengths (blueshifted). These shifts can be measured and translated into a velocity of the hydrogen clouds in space, each of which produces its own spectral pattern. The velocity, in turn, can be translated into a distance, provided some general picture is available which tells the astronomer how distant gas in the galaxy is expected to move. Such information comes from related studies of the motion of stars whose distances are known. The observations of interstellar hydrogen therefore reveal the distribution of the gas in three-dimensional space—two dimensions of position (the two coordinates on the sky) and one of velocity (based on the red- or blueshift observed), which gives the distance.

Researchers who observe HI become accustomed to visualizing their data, such as maps like that shown in Figure 9.2, in three-dimensional space, even if one dimension is, strictly speaking, velocity. They learn to mentally translate velocity into a distance.

Thousands of maps like Figure 9.2 have been produced and published in catalog form. Interpretation of such maps becomes relatively easy once a few general rules are followed. For example, the contour ridge seen passing vertically through the map, around zero velocity, is due to hydrogen gas relatively close to the sun, gas which suffers no significant, systematic Doppler shift. On the other hand, the large structure in the middle of the right side of the map, around velocity +60 km/s, is due to hydrogen clouds in a distant spiral arm located about 30,000 light-years from the sun.

In Figure 9.2, the elliptically shaped contours around +60 km/s are produced by gas which extends between +1° and −1° in latitude, or about 1000 light-years across. This is interpreted as the thickness of the hydrogen layer in that direction and is fairly typical for most of the inner parts of the galaxy.

The negative-velocity hydrogen in Figure 9.2 lies on the far side of the galaxy. The gas at −60 km/s is 60,000 light-years from the sun and is seen at positive latitudes, which is due to an upward warp of the otherwise flat disk of the galaxy in its outer regions.


## Cloud Properties

Radio observations of HI clouds give enormous amounts of information on the properties of individual clouds. In Figure 9.2 a small cloud is visible at −65 km/s and latitude −10′ (10 minutes of arc below the galactic plane). This cloud could be mapped by obtaining more data at adjoining latitudes. In this map the cloud size appears to be about 10 minutes of arc, which translates into a linear size of about 175 light-years across (given that it is located 60,000 light-years away). HI clouds are typically between a couple of light-years, for dense pockets of gas, to a few hundred light-years in extent.

The total number of atoms contributing to an observed spectral line of HI emission can be found by measuring the area under the spectral line. If the diameter of the cloud can be found, this allows the total number of atoms in the cloud to be converted to a density in the cloud. Typical interstellar hydrogen cloud densities are between 3 and 100 atoms per cubic centimeter of space. A spherical HI cloud, two light-years across, whose density is 100 atoms per cubic centimeter, contains a total of $10^{58}$ atoms. The mass of such a cloud would be about 10 solar masses.

## Interstellar Magnetic Fields

If an HI cloud is permeated by a magnetic field the motion of a spinning electron is altered very minutely. The spectral line is split into two lines whose difference in frequency is extremely small but can be measured. In 1959 it was suggested that this phenomenon, known as the Zeeman effect, could be used as a tool for directly determining the strength of the interstellar magnetic field. In 1968, the author of this book first detected the interstellar magnetic field in several HI clouds in the direction of the Cassiopeia A and Taurus A radio sources. Field strengths range from 2 to 20 microgauss (a microgauss is one-millionth of a gauss, the unit by which magnetic fields are measured). By comparison, the earth's magnetic field is about a tenth of a gauss. It is remarkable that radio telescopes can be used to measure not only the temperature and densities in distant clouds, but also the strength of a magnetic field of microgauss (equal to one hundred-thousandth of the earth's field) in a hydrogen cloud 10,000 light-years away.

## Galactic Structure

Extensive interpretation of thousands of hydrogen line contour maps, such as Figure 9.2, can lead to a picture of the large-scale spiral structure of the Milky Way. Because we are located inside the galaxy and stars are mostly obscured by interstellar dust in the galactic plane, it is nearly impossible to see the spiral structure of our galaxy in the distribution of stars, mainly because the stars closest to the galactic plane, which would exhibit this structure, are hidden behind the dust. HI spectral lines, being radio waves, are unaffected by dust, and so the radio astronomer can see through the entire depth of the galaxy. Using maps such as Figure 9.2, the spectral lines due to gas in distant spiral arms are first identified, the velocities converted to distances, and then the location of the main features in the data are plotted on a plan view of the

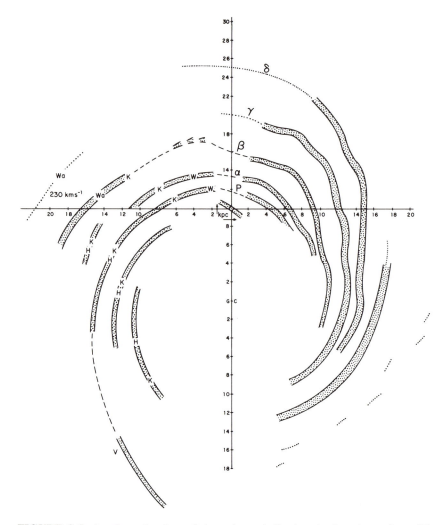

**FIGURE 9.3.** A schematic view of the galaxy, indicating the locations of possible spiral arms as recognized from 21-cm neutral hydrogen data. The sun is located where the two straight lines intersect. The lettering refers to spiral arms labeled by various radio astronomers during their research. The scale of the map is given in parsecs. One parsec is approximately 3 light-years. It is very difficult to infer where spiral arms between the sun and the galactic center are located, so they are not sketched in.

galaxy. An example is shown in Figure 9.3. This process is actually far more complex than here implied, and the study of galactic structure has taken many years of work on the part of dozens of radio astronomers. Figure 9.3 represents but one of the many attempts to picture the spiral structure of the Milky Way.

# The High-Velocity Cloud Mystery

Normal galactic hydrogen will show Doppler shifts of up to 220 km/s, which is the velocity of the sun in its orbit about the galactic center. Distant gas, at rest in space, would show a Doppler shift not greater than would be produced by the motion of the sun, and hence the earth, through space, that is, 220 km/s. Because the galaxy is thin and flat, the HI close to the sun will show only a small velocity, at most 20 km/s, due to random cloud motions. Gas close to the galactic plane, located in distant spiral arms and also partaking of galactic rotation, would be observed to have higher velocities, somewhere between 20 km/s, and the largest velocity possible, 220 km/s. This is the velocity that would be manifested by very distant matter at rest with respect to the center of the galaxy and is determined by the motion of the sun about the galactic center. However, there exist a considerable number of clouds with velocities beyond the values expected from galactic rotation effects. These are known as the *high-velocity clouds* (*HVCs*) and exist mostly in the northern skies, well away from the disk of the galaxy. They show velocities of up to 200 km/s toward the sun while a few in the southern heavens show velocities up to 460 km/s.

Many theories have been proposed for the HVCs, including that they are intergalactic clouds, more nearby gas falling in toward the Milky Way, parts of supernova shells, or members of very distant spiral arms. The latter model would place them at the upper right-hand edge of the pattern in Figure 9.3. The very highest velocity clouds may be located about 100,000 light-years away, in intergalactic space, in a region of sky which seems to be populated with debris left over from a close orbital encounter between the Magellanic Clouds (two galaxies about 150,000 light-years away in the Local Group) and our galaxy. This encounter also produced a 100,000-light-years-long streamer of HI gas, known as the Magellanic Stream, found near the south pole of the galaxy which stretches out of the Magellanic Clouds in a sort of intergalactic arch toward our galaxy.

# Hydrogen in Other Galaxies

Neutral hydrogen in other galaxies has been studied with very high-resolution radio telescopes, including the VLA and the Westerbork radio telescope in the Netherlands. The maps of the HI in those galaxies reveal an amazing variety of structures.

Figure 9.4 shows the HI in the galaxy M81. Spiral patterns can be seen. The hydrogen in our neighbor, M31 in Andromeda, on the other hand, shows little obvious spiral structure (Figure 9.5). However, this HI map looks almost exactly like the map of the continuum radio emission from M31 as a whole. This is shown in Figure 9.6. The synchrotron radiation appears to come from just the same ringlike structure in which the HI is found. In addition, Figure

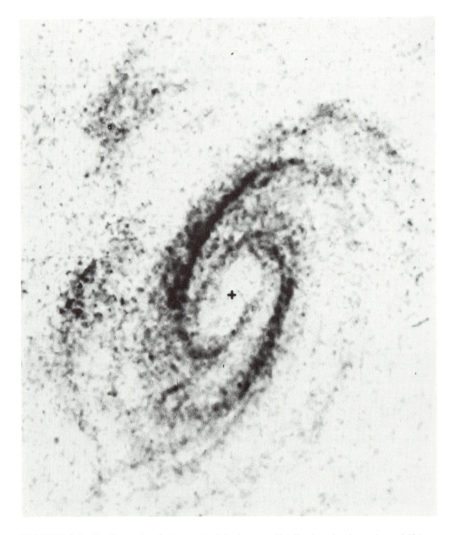

**FIGURE 9.4.** Radiograph of the neutral hydrogen distribution in the galaxy M81, a prototype of a perfect spiral galaxy. Radio emission comes from cosmic rays interacting with magnetic fields in the spiral arms, and a large number of small pointlike sources can be seen. These are either supernova remnants or emission nebulae. Vertical size is 40′. (A. H. Rots, Leiden Observatory.)

9.6 shows that a prominent radio source is located at the center of M31, a source probably very much like the Sgr A source at the center of the Milky Way.

Figure 9.7 is a radiograph of the HI in the galaxy M33, another spiral galaxy in the Local Group, and it reveals anything but a clear spiral pattern.

The study of HI in other galaxies reveals that spiral structure, when it is

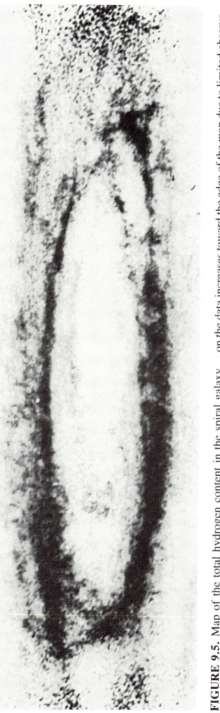

**FIGURE 9.5.** Map of the total hydrogen content in the spiral galaxy M31, located about $2 \times 10^6$ light-years away in the constellation Andromeda, made with a resolution of $24 \times 36$ seconds of arc with the Westerbork Synthesis Radio Telescope in the Netherlands. The noise on the data increases toward the edge of the map due to limited observations in those regions. Vertical size is 40′. (E. Brinks, Leiden Observatory.)

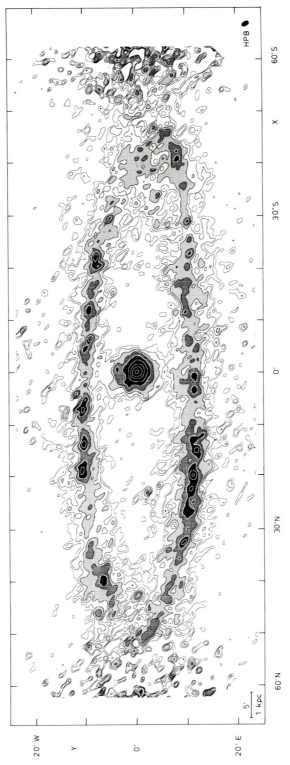

**FIGURE 9.6.** Map of the radio (continuum) emission from M31 at 21 cm observed away from the hydrogen-line wavelength. The map was made with the Westerbork Synthesis Radio Telescope and has a resolution of 54 × 82 arcseconds. The regions of strong radio emission indicate where interstellar magnetic fields in M31 are being illuminated by cosmic rays. Both the continuum emission and the neutral hydrogen gas (Figure 9.5) are concentrated in a ring of radius 30,000 light-years about the center of M31. Note the radio source at the center, which corresponds to a similar source at the center of the Milky Way. The ring coincides with maxima in the distribution of young stars, HII regions, and interstellar dust. (R. Walterbos, Leiden Observatory.)

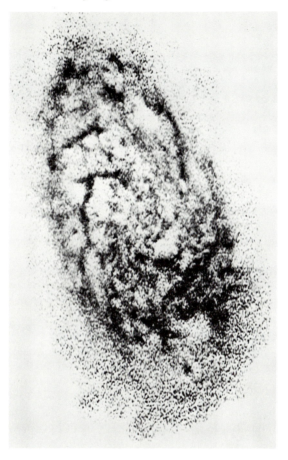

**FIGURE 9.7.** Map of the total hydrogen content of the spiral galaxy M33 obtained with the Westerbork Synthesis Radio Telescope, set to produce a resolution of 12 × 24 arcseconds. There is only the vaguest suggestion of spiral patterns in the distribution of hydrogen in this galaxy. The blurring around the edge of the map is produced in regions where noise in the radio receiver was comparable with the weak signals from M33. Vertical size is 1° 20′. (E. Deul, Leiden Observatory.)

present, is usually not simple, and often is hard to find, even in galaxies which are obvious spirals when viewed optically. The theory of why galaxies show spiral structure, although highly developed, seems to be ignored by many galaxies!

HI studies have given us an enormous amount of information regarding the physical conditions in interstellar space, data which can then be compared with infrared, optical, and ultraviolet observations. Hydrogen gas is the most basic substance in the universe, and in our galaxy the space in and around the hydrogen clouds contains far more than just hydrogen. It is there that we find the interstellar molecules.

# 10

# Interstellar Molecules

## Chemical Factories in Space

Alcohol, ether, water, ammonia, carbon monoxide, acetylene, and embalming fluid are some of the over 60 molecular species so far discovered in vast interstellar clouds of dust and gas. Most of the recently discovered interstellar molecules, listed in Table 10.1, are carbon-based—that is, organic molecules. The chemistry of this type of molecule is involved in all life on earth. Twenty years ago no one would have predicted that space was replete with many of the same molecular structures which are essential for the origin and evolution of life.

There are several striking things about the contents of Table 10.1. Apart from its challenge to memory and spelling ability, the predominance of carbon (C), hydrogen (H), nitrogen (N), and oxygen (O) in the molecules is fascinating. These four atoms make up the simple alphabet of life. With the four ''letters'' C, N, O, and H and a few punctuations provided by atoms such as phosphorus (P) you can build words (molecules) and combine them into sentences, such as protein or DNA molecules, and into entire chapters and books which are living organisms.

The contents of Table 10.1 is our first hint that if there is life elsewhere in the galaxy (or the universe, because several of these molecules have been found in other galaxies as well), its chemistry is going to be similar to our own.

The interstellar molecules are found either in dense interstellar dust clouds or in shells around stars. Two of the most prominent molecular clouds lie in the direction of Sagittarius (in the cloud Sgr B2, which is located near the galactic center) and in the immediate vicinity of the Orion nebula. Virtually all the complex species in Table 10.1 are found in either one or both of these clouds, not because these are the only two in the galaxy, but because they are, respectively, the largest known cloud (Sgr B2) and the one closest to us (Orion, 1200 light-years away). Typical interstellar chemical factories are found around HII regions or emission nebulae, and Figure 10.1 shows an example of such an emission nebula, M16. Dust clouds, seen in silhouette against the emission from the luminous gas, are the birthplaces of molecules and certainly

**FIGURE 10.1.** Optical image of the emission nebula (HII region) known as M16 in the constellation of Serpens. Hot gas radiates light, heat, and radio waves. Dust clouds in and around the nebula obscure the light from more distant stars. The bright edges along some of the dust clouds indicate regions where gas is erupting from the cloud surfaces. (National Optical Astronomy Observatories.)

**TABLE 10.1.** Table of interstellar molecules.

| Chemical symbol | Name of molecule | Year of discovery | Part of spectrum |
|---|---|---|---|
| *Two-atom molecules* | | | |
| CH | Methyladine | 1937 | Visible |
| CN | Cyanogen | 1940 | Visible |
| CH$^+$ | Methyladyne | 1941 | Visible |
| OH | Hydroxyl | 1963 | Radio 18 cm |
| CO | Carbon monoxide | 1970 | Radio 2.6 mm |
| H$_2$ | Molecular hydrogen | 1970 | Ultraviolet |
| CS | Carbon monosulfide | 1971 | Radio 2.0 mm |
| SiO | Silicon monoxide | 1971 | Radio 2.3 mm |
| SO | Sulfur monoxide | 1973 | Radio 3.0 mm |
| NS | Nitrogen sulfide | 1975 | Radio 2.6 mm |
| SiS | Silicon monosulfide | 1975 | Radio 2.8 mm |
| C$_2$ | Diatomic carbon | 1977 | Infrared |
| NO | Nitric oxide | 1978 | Radio 2.0 mm |
| HCl | Hydrogen chloride | 1985 | Infrared |
| *Three-atom molecules* | | | |
| H$_2$O | Water | 1968 | Radio 1.4 cm |
| HCO$^+$ | Formyl ion | 1970 | Radio 3.4 mm |
| HCN | Hydrogen cyanide | 1970 | Radio 3.4 mm |
| HNC | Hydrogen isocyanide | 1971 | Radio 3.3 mm |
| OCS | Carbonyl sulfide | 1971 | Radio 2.7 mm |
| H$_2$S | Hydrogen sulfide | 1972 | Radio 1.8 mm |
| C$_2$H | Ethynyl radical | 1974 | Radio 3.4 mm |
| N$_2$H$^+$ | Diazenylium | 1974 | Radio 3.2 mm |
| HCO | Formyl radical | 1975 | Radio 3.5 mm |
| SO$_2$ | Sulfur dioxide | 1975 | Radio 3.6 mm |
| HCS$^+$ | Thioformylium | 1980 | Radio 3 mm |
| SiC$_2$ | Silcyclopropyne | 1984 | Radio 1–3 mm |
| *Four-atom molecules* | | | |
| NH$_3$ | Ammonia | 1968 | Radio 1.3 cm |
| H$_2$CO | Formaldehyde | 1969 | Radio 6.2 cm |
| HNCO | Isocyanic acid | 1971 | Radio 3.4 mm |
| H$_2$CS | Thioformaldehyde | 1971 | Radio 9.5 cm |
| C$_2$H$_2$ | Acetylene | 1976 | Infrared |
| C$_3$N | Cyanoethynyl radical | 1976 | Radio 3.4 mm |
| HNCS | Isothiocyanic acid | 1979 | Radio 3 mm |
| HOCO$^+$ | Protonated carbon dioxide | 1980 | Radio 3 mm |
| HCNH$^+$ | Protonated hydrogen cyanide | 1984 | Radio 2 to 3 mm |
| C$_3$H | Proponyl radical | 1984 | Radio 3 mm |
| C$_3$O | Tricarbon monoxide | 1984 | Radio 1.7 cm |
| HCNH$^{2+}$ | Protonated hydrogen cyanide | 1985 | Radio 4 mm |
| *Five-atom molecules* | | | |
| HCOOH | Formic acid | 1970 | Radio 18 cm |
| HC$_3$N | Cyanoacetylene | 1970 | Radio 3.3 cm |
| CH$_2$NH | Methanimine | 1972 | Radio 5.7 cm |
| NH$_2$CN | Cyanamide | 1975 | Radio 3.7 mm |
| CH$_2$CO | Ketene | 1976 | Radio 2.9 mm |
| C$_4$H | Butadiynyl radical | 1978 | Radio 2.6 mm |

**TABLE 10.1.** (*continued*)

| Chemical symbol | Name of molecule | Year of discovery | Part of spectrum |
|---|---|---|---|
| SiH$_4$ | Silane | 1984 | Infrared |
| C$_3$H$_2$ | Cyclopropenylidene radical | 1985 | Radio 2 cm |
| *Six-atom molecules* | | | |
| CH$_3$OH | Methyl (wood) alcohol | 1970 | Radio 36 cm |
| CH$_3$CN | Methyl cyanide | 1971 | Radio 2.7 mm |
| NH$_2$CHO | Formamide | 1971 | Radio 6.5 cm |
| CH$_3$SH | Methyl mercaptan | 1979 | Radio 3.0 mm |
| *Seven-atom molecules* | | | |
| CH$_3$C$_2$H | Methylacetylene | 1971 | Radio 3.5 mm |
| CH$_3$CHO | Acetaldehyde | 1971 | Radio 28 cm |
| CH$_3$NH$_2$ | Methylamine | 1974 | Radio 3.5 mm |
| CH$_2$CHCN | Vinyl cyanide | 1975 | Radio 22 cm |
| HC$_5$N | Cyanodiacetylene | 1976 | Radio 3.0 cm |
| *Eight-atom molecules* | | | |
| HCOOCH$_3$ | Methyl formate | 1975 | Radio 18 cm |
| CH$_3$C$_3$N | Methyl cyanoacetylene | 1983 | Radio 1.5 cm |
| *Nine-atom molecules* | | | |
| CH$_3$CH$_2$OH | Ethyl alcohol | 1974 | Radio 2.9 mm |
| (CH$_3$)$_2$O | Dimethyl ether | 1974 | Radio 9.6 mm |
| C$_2$H$_5$CN | Ethyl cyanide | 1977 | Radio 3 mm |
| HC$_7$N | Cyanotriacetylene | 1977 | Radio 2.9 cm |
| CH$_3$C$_4$H | Methyl diacetylene | 1984 | Radio 1.5 cm |
| *Eleven-atom molecules* | | | |
| HC$_9$N | Cyanoctatetrayne | 1977 | Radio 2.9 cm |
| *Thirteen-atom molecules* | | | |
| HC$_{11}$N | Cyanodecapentyne | 1981 | Radio 1.3 cm |

the storehouses as well, because there the molecules are protected from the disruptive effects of starlight.

Thousands of molecule-bearing clouds exist in the Milky Way, and the most pervasive constituents of these clouds are carbon monoxide (CO), seen in virtually all directions, and ammonia (NH$_3$). Molecular hydrogen is present in the largest amounts, but it is extremely difficult to observe (see below). Formaldehyde (H$_2$CO = embalming fluid) is also extremely widespread, although formic acid (HCOOH), the substance that gives ants a remarkably acrid taste, is less common.

Table 10.1 does not list the isotopic variations of many of these molecules. An isotope of an atom is one which contains an abnormal number of neutrons in its nucleus. For example, the carbon atom usually contains a total of 12 particles at its nucleus, six protons and six neutrons. In this state it is known as carbon-12. If it should have an extra neutron in its nucleus it is called carbon-13, which is an *isotope* of carbon. At least 50 variations of the molecules in Table 10.1, mostly involving isotopes of carbon and oxygen, have been observed.

## What Is a Molecule?

When two hydrogen atoms are brought very close together, they lock in an embrace enforced by the cohesive power of their electrons, which orbit both protons. The electrons act as a glue that holds the two hydrogen atoms together to form the molecule. In large molecules dozens of electrons may be involved in the process of herding (or bonding) together dozens of atoms into stable flocks.

Molecules may be destroyed by giving them too much energy. For example, excessive heat or UV radiation causes the individual atoms to tear themselves loose from their partners. When the adhesive power of the electrons is overcome, the molecule breaks apart.

In interstellar space there are several ways to form molecules and many ways to destroy them. The fact that molecular species are found in space at all means that the formation process is generally more effective than destruction; otherwise there would be no molecules to be observed! Apparently dust in the interstellar clouds acts to protect the molecules from disruptive stellar energies, especially ultraviolet radiation.

The combination of two hydrogen atoms and an oxygen atom produces the water molecule, a most basic molecule for living things. Interstellar clouds contain vast amounts of water, whose spectral line was first observed in 1968.

## Molecular Spectral Lines

Interstellar molecules are recognized because they emit radio waves that we observe in the form of spectral lines. An example is shown in Figure 10.2, the spectral line due to carbon monoxide as observed by a radio telescope pointed in the direction of the Sagittarius radio source at the galactic center.

Most interstellar molecules are asymmetrical in shape. For example, an oxygen atom and a hydrogen atom combine in a molecule known as hydroxyl (OH), shaped something like a dumbbell, with the large oxygen and small hydrogen atoms glued together by an encircling electron that originally belonged to the hydrogen. Such a molecule is capable of rotating in two ways; either end-over-end or around an axis drawn between the two atoms. Whenever two or more energy states are possible, transitions occur between them, and that means that emission of energy at specific signature wavelengths, as was discussed in the previous chapter, is possible. Some molecular species have dozens of possible energy states, depending on their architecture. The signature spectrum of such

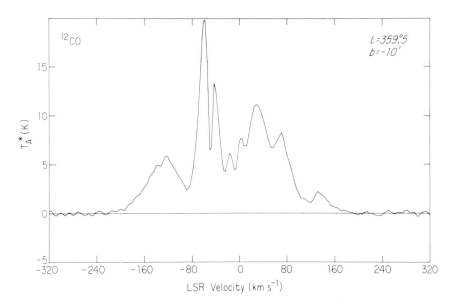

**FIGURE 10.2.** A typical emission spectrum produced by an interstellar molecule—in this case, carbon monoxide—observed in a direction close to the galactic center. The many peaks indicate that several well-separated clouds are the cause of the spectral peaks. Intensity is given as a temperature measured at the antenna and the velocities are in terms of the radio astronomical system defined in the Appendix. (T. M. Bania.)

molecules may contain hundreds of individual "lines" located over a wide range of wavelength, either optical, radio, infrared, or ultraviolet.

The most important interstellar molecule is molecular hydrogen ($H_2$), believed to constitute over 50% of the molecular mass in the galaxy. Because the molecule is symmetric (being made of two identical hydrogen atoms), it has no differentiated energy states between which transitions leading to radio-wavelength spectral lines can occur. It is observed at IR wavelengths, although its presence is usually inferred from the widespread distribution of CO, which is only formed in dust clouds containing a lot of molecular hydrogen.

The molecules listed in Table 10.1 are identified on the basis of spectral lines whose frequencies have been measured in laboratory experiments. In addition, about 150 spectral lines of unknown origin, due to possibly 50 other complex organic molecules, have been discovered by radio astronomers.

Early in 1985 several of the previously unidentified lines were associated with a ring molecule, $C_3H_2$, and this may soon turn out to be the most important molecule yet discovered in space, not only because it is present in significant amounts in virtually every location known to contain interstellar molecules, but because ring molecules are important to life chemistry.

Conditions in space are so vastly different from those on earth that it is incredible that we should find so many substances in vast interstellar clouds which also are common on our planet. Even though temperatures in space are close to absolute zero and the densities of materials are far below those in our atmosphere, common molecules like water, alcohol, carbon monoxide, and ammonia exist out there.

Table 10.1 contains the variety it does because those are the molecular types about which enough was known from lab work to allow their identification from spectral signatures observed by radio astronomers. They are also those molecules which could be observed from earth because their short-wavelength radio signals reached the earth without being absorbed in the atmosphere.

## Millimeter-Wave Windows

Water and oxygen in the atmosphere turn the sky opaque below a radio wavelength of about 1 cm. However, there are several "windows" where the transparency at certain wavelengths is still fairly good. It is possible to squint through these windows, especially if the radio telescope is located above much of the water, such as on a mountain top or in a dry desert. The highly accurate 12-m telescope of the NRAO (Figure 10.3) on Kitt Peak near Tucson, Arizona, is well situated to study molecular line emission, especially around millimeter wavelengths. The existence of a limited number of windows has meant that many of the molecules detected to date have been selected because they happen to shine through these windows and because radio telescopes like the 12-m are available for the search.

**FIGURE 10.3.** The 12-m radio telescope of the National Radio Astronomy Observatory, designed to operate at very short wavelengths of a few millimeters. In order to make the best use of atmospheric "windows" which allow some of the millimeter waves, otherwise completely absorbed by water vapor, to reach the ground, the telescope is situated at an elevation of 6300 feet, on Kitt Peak, west of Tucson, Arizona. The antenna is housed in a protective dome which shields it from excessive temperature variations and wind. (National Radio Astronomy Observatory.)

# Masers in Space

In 1965, three groups of observers in the USA, studying OH emission in the direction of HII regions, discovered a spectral line at 1665 MHz which was completely unexpected. Although this was one of the frequencies at which interstellar OH was expected to radiate, OH should have been seen at three other frequencies (1612, 1667, and 1720 MHz) as well, but at first those signals were not found. The radio astronomers had, in fact, discovered a "maser" in space.

The acronym MASER refers to "microwave amplification by stimulated emission of radiation." Microwave amplification refers to the amplification of waves at short radio wavelengths, known as microwaves. Stimulated emission of radiation refers to an interesting phenomenon. Under certain conditions molecules may emit far more energy than expected, provided energy is pumped into them by some external energy source. The OH molecules in clouds around an HII region, for example, can absorb light from very red stars, and that light pumps the molecules into excited states. The added energy can be radiated away at a particular spectral line frequency. Other interstellar molecules showing the maser effect include water, silicon monoxide (SiO), formaldehyde, and methyl alcohol.

About 200 water maser sources are found in regions of star formation (HII regions). Others are associated with old, highly evolved stars entering their dotage. In either case they appear to be associated with clouds where the densities range from $10^5$ to as high as $10^{11}$ particles per cubic centimeter, about as dense as anything yet discovered in space. To make the maser work, the pump source has to be about 10,000 times more luminous than the sun. This condition can be met by newly formed massive stars or clusters of such stars. The typical water spectral line shows many components (Figure 10.4) and the velocity spread, or Doppler shifts, inferred from such spectra reveals that material in the clouds is in motion. The intensity of parts of the spectrum, which may vary from day to day, is very great. Intensity is given in "flux units" (now called janskies). The water maser is, within its narrow lines, really the brightest radio source in the sky. For comparison, the strongest radio source in the sky, Cassiopeia A, produces only about 1000 flux units and the brightness of typical radio sources whose radiographs are shown in this book is in the range from a hundredth to several flux units.

In the direction of HII regions, such as the Orion nebula, the water masers appear in clusters and several small clusters are spread over the area of the nebula. It is believed that each small group of water sources, called a "center of activity," may be due to maser amplification within the envelope, or cocoon, surrounding a specific star. Several such stars may lie within the HII region, or very close to its boundary.

Other masers are associated with variable stars, which show SiO, water, and OH masers. These stars are known to have molecule-bearing circumstellar envelopes which are being ejected in a relatively orderly manner, unlike the violent phenomena observed in novae or supernovae. Molecules in these enve-

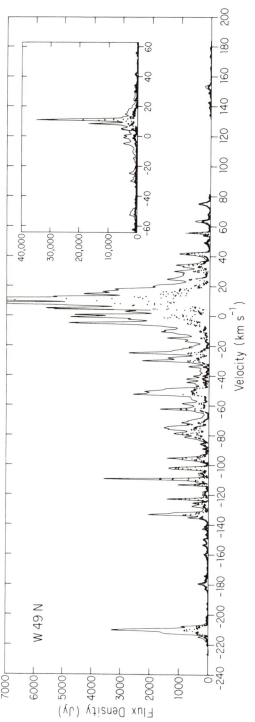

**FIGURE 10.4.** A spectrum of the water maser emission from the northern part of the HII region W49 (Figure 8.13), observed in June 1978 with a very high-resolution radio telescope known as the VLBI (see Chapter 17). Such water masers are highly time variable. The intensity of the lines is enormous (see text) and the separate peaks indicate individual maser sources in the vicinity of W49. The inset shows the central portion of the spectrum on a ten times larger scale. The solid line is the total intensity of the emission at the time of observation and the dots refer to high-resolution observations which were sensitive to only some of the radiation. (R. C. Walker, D. N. Matsakis, and J. A. Garcia-Barreto.)

lopes are pumped by collisional processes in these gases as they expand away from the star.

In a typical stellar maser source, such as VX Sgr, shown in Figure 10.5, the SiO maser is found close to the surface of the star, the water somewhat farther out, in the expanding circumstellar envelope, and the OH even farther out, where the envelope makes contact with surrounding interstellar material.

Figure 10.6 is another fascinating example of how clearly radio astronomers have learned to see in recent years. This figure shows a shell of OH gas expanding away from a star. The OH is embedded in lots of dust, probably formed right at the star. The source shown in the figure is located at galactic longitude 127.8° and latitude 0°, and the coordinates are used to name the object, that is, OH 127.8–0.0. This maser source is associated with a star which is invisible behind a shell of dust. The position coincides with a source of infrared radiation and the maser emission is most likely being produced in a shell of dust and gas being ejected from the invisible star, very likely a giant star in its last phases of life. This is dramatic evidence for expansion of the shell away from

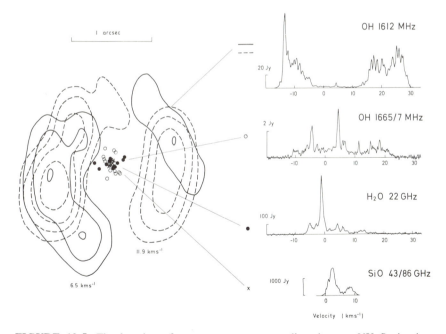

**FIGURE 10.5.** The location of maser sources surrounding the star VX Sagittarius. The silicon monoxide (SiO) masers are closest to the central star followed by the water (H$_2$O) and OH masers. Surrounding the star is an envelope of material in which another OH maser radiates. Two distinct shells of gas (solid and dashed contours) due to gas at slightly different velocities are found in the 1612-MHz OH maser lines. (R. J. Cohen, Nuffield Radio Astronomy Laboratories.)

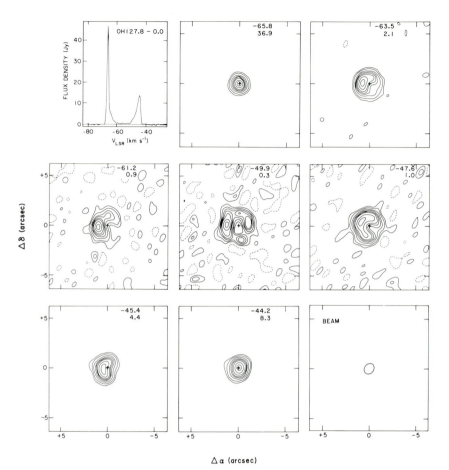

**FIGURE 10.6.** The circumstellar shell discovered around the source known as OH 127.8–0.0. This is a star which emits strong infrared and OH maser signals. The numbers at the top of each frame indicate the velocity of the map. At the velocity of $-49.9$ km/s (the negative sign indicates motion toward us) the shell or ring around the star is dramatically evident. The near and far sides of the expanding shell are seen at the two end frames at $-65.8$ and $-44.2$ km/s. The spectral line produced by the entire object is shown in the upper left frame. This double-peaked spectrum is now known to be characteristic of such stars surrounded by expanding envelopes of gas. In this case the expansion is at about 13 km/s. (P. F. Bowers, K. J. Johnston, and J. H. Spencer.)

the star at 13 km/s. At the two extreme velocities of the OH maser spectrum, which has the shape indicated in the top left-hand frame, the source appears like a point, indicated by the concentric contours. But at velocities in between, the maser looks like a ring. These maps are consistent with the source being an expanding shell. Many OH masers have been found to show these sorts of patterns. Masers are not only associated with giant stars at the end of their lives, but also with very young stars.

## Giant Molecular Clouds

Giant molecular clouds (GMCs) are the most massive objects (up to 10 million solar masses) in the galaxy, consisting almost entirely of molecular hydrogen and carbon monoacide (CO) and riddled through with many of the molecules listed in Table 10.1. GMCs were discovered because they contain enormous quantities of CO. A GMC, which is usually surrounded by an enveloping cloud of atomic hydrogen, gas which is absent inside the cloud, is typically 150–250 light-years in diameter. It contains one or more dense cores a few dozen light-years across.

GMCs are found along the Milky Way, and Figure 10.7 shows a section of sky where the CO clouds have been mapped and compared with the dust seen against distant stars. The outlines of the CO cloud agree perfectly with the dust.

The majority of the GMCs, 4000 of which may exist in the galaxy, are found between 12,000 and 24,000 light-years from the galactic center. A GMC in Sagittarius is one of the most dramatic objects in the galaxy and contains 3–5 million solar masses of mostly molecular hydrogen.

The GMCs are clearly the birthplaces for stars. Evidence for star formation in these clouds comes from the presence of bright infrared sources and masers within the cloud boundaries. The observed HII regions, which often show hot matter streaming into space, may be produced by stars recently formed near the edge of GMCs. These stars then eat their way into the surrounding molecular hydrogen, destroy it, and ionize the atomic (neutral) hydrogen so produced. Ionized gas streams away from the HII region, which appears like a blister at the surface of the GMCs.

Figure 10.8 is a map of the northern part of the HII region known as W49. (See Figure 8.13 for a radiograph of W49.) It is the largest such nebula in the galaxy, but is heavily obscured by dust clouds. The radio map shows seven distinct sources (marked A–G) of radio emission, each believed to be a concentrated HII region surrounding a very young star. These sources are less than 0.01 light-years across. The one labeled G looks like a ring, suggesting the presence of a shell whose diameter is about 0.1 light-years. There may be two shells present, one of which includes the sources labeled G2 and G3.

This map reveals further remarkable properties related to the maser sources discussed above. The OH maser locations are shown on the map in Figure 10.8 and about a dozen water maser sources are found just below the peaks labeled G2 and G3. The masers are all located close to the young, but invisible, stars.

The total energy of the stars required to create the compact HII regions is a million times that produced by the sun, yet only seven stars may be involved.

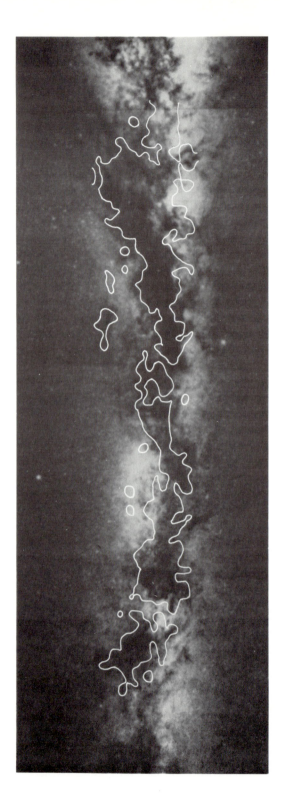

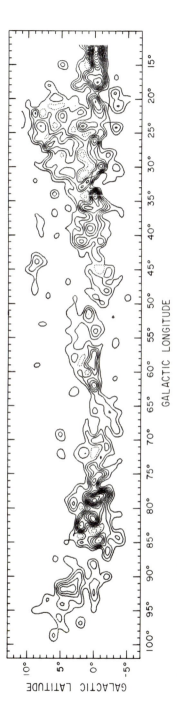

**FIGURE 10.7.** The location of CO clouds plotted as a function of galactic coordinates (lower frame) and the outlines sketched onto an optical photograph of the Milky Way (upper frame) between Sagittarius and Cassiopeia. The photomosaic is a composite of four photographs obtained by W. C. Miller with the Mt. Wilson telescope. The CO is clearly associated with the dust clouds which obscure more distant starlight. (P. Thaddeus.)

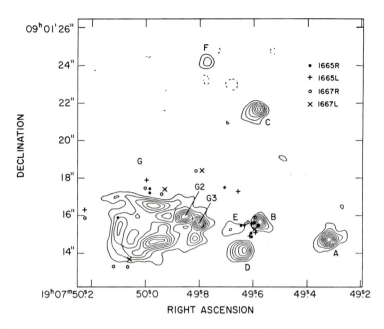

**FIGURE 10.8.** The location of masers in the HII region W49. This map was made at 2-cm wavelength with the VLA and shows seven small sources of intense radio emission (concentric contours) in the northern part of the nebula, and the location of the OH masers associated with them is indicated. These sources are, in turn, associated with the central hot spots seen in the radiograph of this emission region, Figure 8.13. (Observ-ers—J. W. Dreher, K. J. Johnston, W. J. Welch, and R. C. Walker.)

Furthermore, the OH masers indicate a flow of material falling in toward one of the compact components.

## The Stages Immediately Following Star Birth

In regions of active star formation several other phenomena have recently been observed, again through studying the molecular line emission, which indicate that stars just about to start shining eject a lot of material at hundreds of km/s. These objects are known as T Tauri stars, named after a variable star discovered in the constellation of Taurus as long ago as last century. There are many T Tauri stars and they share a common property that immediately surrounding them, within an arcminute or so, a small nebula can be seen. This appears to be interstellar dust and gas immediately around the protostellar object. Gas seems to be moving both in and out of the T Tauri star while, incredibly, at some distance from these stars other small nebulae are found streaming away from the T Tauri star in much the same way that a jet in a radio galaxy is pointed away from its nucleus. These other nebulae are known as Herbig–Haro objects (HH objects), after their discoverers. The HH objects were for a very long time a mystery because they did not appear to have any stellar objects associated with them. The stellar objects are not located in the HH objects, but are light-years away (see Figure 10.9). The T Tauri stars which have ejected material, including large numbers of molecules, produce the HH objects.

The ejecta from the T Tauri are called bipolar flows. Such bipolar flows,

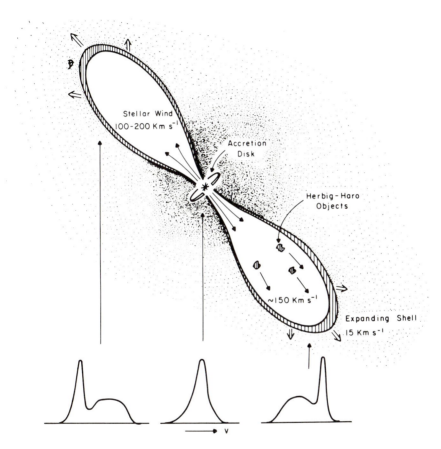

**FIGURE 10.9.** A schematic representation of the molecular cloud known as L1551, showing two lobes extending 1.5 light-years in opposite directions from an IR source buried within the cloud. One lobe is associated with seven Herbig–Haro objects, small nebular patches of radiating gas. The CO emission in the double-lobed structure arises from a dense shell of material swept up by the stellar wind rushing out at 200 km/s. This wind emerges from the central hole of an accretion disk and moves into space, piling up matter ahead of it as its boundary expands slowly, at about 15 km/s. The shape of the CO spectral line at different points is consistent with observations of this source. Other bipolar flows are believed to be very similar to this well-studied case. (R. Snell, R. B. Loren, and R. L. Plambeck.)

observed by means of the molecules they contain, have been found around other species of stars as well. In the case of the T Tauri stars, light emission is produced at the shocked region where the ejecta push up against the surrounding molecular cloud.

It appears that when a T Tauri object, soon ready to turn on its nuclear furnace, begins to stir in its cocoon the energy it generates, due to the collapse from dimensions of several light-years across to stellar size, is so great that it can hurtle a great amount of matter outward (a millionth of a solar mass per year is typical). This matter travels in two directions for the same reason that radio sources emit jets in two directions—something prevents the material from moving in the other directions. That something is an accretion disk around the star, formed by matter that has most recently gathered in an attempt to become part of the star. These gases are pulled into an accretion disk just as a black hole sucks in matter in a galactic core, but this time the gas has a more benign fate awaiting it: If it does not get tossed back out again or fall into the star, this material will subsequently form planets.

A remarkable close-up of a bipolar flow is shown in Figure 10.10. This radiograph shows the radio emission from the HII region known as S106, a source of infrared radiation, located about 1200 light-years away. The angular scale on this radiograph is about 1.5 minutes of arc on a side. At first glance the image may look like a confused mess. However, across the middle of this nebula is a dark band, caused by a lack of radio emission. Within that band a tiny bright point shows the radio emission from the surface of the star giving rise to this nebula. The star is surrounded by a flat disk of gas and dust (density $10^6$ cm$^{-3}$) seen nearly edge-on, another example of an accretion disk, this time almost revealed directly. We are seeing its shadow, so to speak. The disk contains 0.01 solar mass of material, and it produces the dark band because it generates only very weak radio waves. Material is streaming away from the star, along the axis of the disk, again something like the radio jets seen in quasars and radio galaxies, but this example of a bipolar flow is not jetlike. Here the hot gas is broadly spread, shows filamentary structure, and is very patchy. Based on molecular line observations (CO, ammonia, CS, and CN), the flow in this nebula is estimated at about 100 km/s. This flow ejects a millionth of a solar mass per year. The material is almost certainly emerging from a very young, relatively massive star (called a BO star) which is just shrugging off the confining influence of its surrounding accretion disk. Later, when the star has settled into adulthood and the strong wind generating the bipolar nebula has died down, material in the disk will probably form planets.

## Molecules and Galactic Structure

Because molecules are so widespread, some of them, in particular CO, have provided a dramatic new way to discover what the spiral structure of the galaxy looks like. CO clouds lie close to the galactic plane and they produce clear

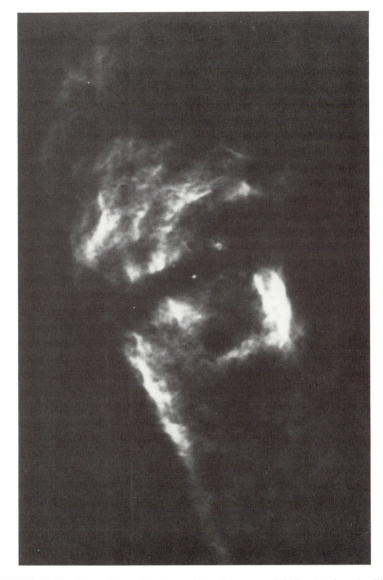

**FIGURE 10.10.** The radiograph of an object known as S106, showing the bipolar flow very dramatically. This 6-cm-wavelength image shows a dark lane across the center of the object. The bright point within that lane is the star responsible for the flow. The accretion disk surrounding the star is invisible, but must be lined up along the dark region, which is due to an absence of radio emission. As the flow of gas emerges from the central hole of the accretion disk (see Figure 10.9), it expands freely into space and produces the double nebula. This object may be the one most likely to represent what occurred in the early formation of the solar system, when the sun was blowing gas out of the regions where planets later formed. Vertical size is 1.5'. (J. Bally, R. L. Snell, and R. Predmore.)

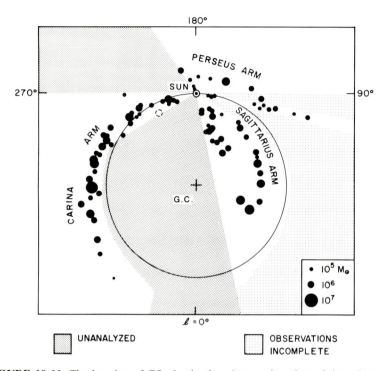

**FIGURE 10.11.** The location of CO clouds plotted on a plan view of the galaxy. The observed velocities of CO emission associated with giant molecular clouds in the direction of the southern constellation of Carina, as well as those associated with many other clouds, have been converted into distances and plotted on this map, which indicates what the galaxy would look like when viewed from a great distance. The CO clouds lie along spiral arms, and comparison with Figure 9.3 shows that the P feature in the latter corresponds to the Perseus arm in this figure. (P. Thaddeus.)

spectral lines whose velocities can be measured—again related to the Doppler shift derived from comparing observed with expected frequencies. Using the same sort of approach as was discussed in Chapter 9 for hydrogen gas, the distance to the CO clouds can be measured. Figure 10.11 shows where many of the GMCs are located. The region of the spiral arm in the constellation of Carina has been particularly well defined. This map may be compared with Figure 9.3.

The study of GMCs has exploded in the last few years, and in the next decade such studies, along with the discovery of more molecules, as well as the soon-to-be-expected boom in observations of the widely spread and easy to observe ring molecule $C_3H_2$, will create both a revolution and an evolution of our ideas of how and where stars form and of the processes that accompany their birth.

# Part IV

## Stellar-Type Radio Sources

# 11

---

# Pulsars

## Expecting the Unexpected

Many of the really dramatic breakthroughs in radio astronomy have occurred because of accidental discovery (sometimes called "serendipity") rather than because the astronomer knew what he or she was looking for! However, it takes more than being at the scene to become a faithful witness to a new phenomenon. It takes an openness to the unexpected as well as a willingness and skill to follow the clues nature provides, wherever they may lead. Only then will the scientist be primed to discover something new upon the face of the sky.

Pioneering discovery usually occurs in a climate of "tradition," in the context of accepted "models" or ideas about the way things are. Within the scientific establishment tradition and authority rule as much as they do in other systems of organized belief. However, the very essence of the scientist's philosophy is built on the awareness that scientific knowledge will change and evolve, despite inertia associated with tradition. Consequently, our fund of knowledge about the universe and our interpretations of natural phenomena systematically move toward new levels of understanding. The potential for progress is built into scientific methodology. Therefore, when a researcher has that peculiar combination of skill, luck, persistence, and a willingness to risk the censure which may accompany the expression of new ideas, there is a chance that he or she may achieve an exciting breakthrough and contribute measurably to real "progress."

Radio astronomy is replete with serendipitous discovery and the pulsar saga is a prime example. Radio waves from *pulsars* (pulsating radio sources) were discovered in 1967, but it was months before the researchers convinced themselves that they hadn't, in fact, picked up radio signals from extraterrestrial civilizations, and so felt free to announce to the world the detection of a new class of radio source in the universe.

## Scintillation of Radio Sources

The pulsar story can be traced back to the mid-1960s, when a pioneering survery in the quest for new radio sources was conducted at Cambridge University. Some of the newly discovered sources (most of which later turned out to be quasars or radio galaxies) seemed to change brightness from minute to minute, but only when they were observed close to the direction of the sun. This phenomenon is called *scintillation* and is produced when the radio waves pass through a patchy cloud of thermal electrons. Such clouds will cause the radio waves to alter their path slightly, jiggling back and forth from minute to minute. This results in the scintillation of the radio source. The reason for this is directly analogous to the explanation for the twinkling of starlight.

The earth's atmosphere contains a lot of water vapor, and at high altitudes starlight is deflected on its path through this vapor. Because of their great distance from us, stars appear as points of light, but when a beam of light from such a point moves through irregularities in the atmosphere the light is deflected back and forth and actually reaches our eyes after having traveled slightly different paths from moment to moment. This effect causes the star to *twinkle*. Planets, however, do not twinkle because they do not appear as points of light. They are so close to us that they appear as small disks and the light reaching our eyes from a planet is the combination of thousands of twinkling beams, each of which impinges on our eye at the same moment. The net effect is that the planet appears to shine steadily. (This classical distinguishing mark—that stars twinkle and planets don't—allow you to tell the difference between a star and a planet on a clear night.)

The smallest-diameter radio sources scintillate when their beams pass through electron clouds, which are continually blown out of the sun in the "solar wind." Larger-diameter radio sources, however, shine steadily. It was realized that observations of radio source scintillation contain information about both the properties of the particle clouds streaming from the sun and the angular size of the radio sources themselves.

After the discovery of radio source scintillation, the Cambridge radio astronomers realized that a good, cheap radio telescope could be built that would allow persistent monitoring of this phenomenon, so that radio source diameters, largely unknown at the time, could be estimated. And so the antenna with which pulsars were discovered came to be built.

## The Discovery of Pulsars

Pulsars were discovered because of remarkable persistence on the part of Jocelyn Bell, a graduate student at Cambridge University in England. The economy-size radio telescope, constructed by loving and eager student labor, consisted of over a thousand wooden posts, each about 10 feet high, with miles of wire strung between them. This telescope was built before computers were as pervasive as they now are so automatic pen recorders were used to display the data. In such a recorder a pen draws a line on a paper chart which is automatically unrolled as the machine creates its recording. Analysis of the radio observations required inspection of these charts and the measurement of deflections from the so-called "baseline," the normal path of the line drawn on the paper in the absence of a radio source in the beam of the antenna. In modern radio observatories the data are directly fed into computers, where the information is lost from sight until the final numbers are printed out. However, the scintillation experiment produced data on 400 feet of chart paper every day.

The antenna was located near Cambridge, England, and plenty of locally produced radio interference (such as automobile ignition) contributed to the deflections of the pen. These deflections appeared similar to those expected from scintillating radio sources, and Bell, studying her ration of 400 feet of paper a day, soon became an expert in recognizing which was which. In the course of her work she experienced what other graduate students assigned the tiresome task of studying endless miles of data sometimes discover—the brain is capable of the most extraordinary feats of memory regarding apparent trivia. She noticed that the recordings showed a faint signal which could not be explained by interference or scintillation or any other natural causes then known to astronomers.

Over the months Bell looked at miles of paper charts and she found that this "little bit of scruff" did not go away, and that it occurred at night when the sun was below the horizon and no scintillating sources were expected. Persistently she monitored the charts and noticed that the "scruff" appeared four minutes earlier each day.

The stars move across the skies at a different rate from that of the sun. Another way of stating this is that the length of the solar day, 24 hours, is different from the length of the day measured with respect to the stars, which is 23 hours and 56 minutes. All the stars thus appear in slightly different directions, as seen with respect to the horizon, from night to night, an effect that is noticeable to even the casual observer over periods of a month or so. A given star appears directly south, say, four minutes earlier every day.

The "bits of scruff" were unlikely to be man-made interference, which would tend to occur either randomly or at the same time each night. Since the time of arrival of the "scruff," of which four sources had been recognized, shifted by about four minutes per day, the signals had to be coming from something associated with the starry heavens. When the research group at Cambridge finally confronted the reality of the discovery, they studied the new radio sources more carefully and were astonished to find that the radio signals were pulsating with impressive regularity, so regular that it required the best available clocks to measure the arrival time of these "pulse trains," an example of which is shown in Figure 11.1. One of the original sources, named CP 1133 (for Cambridge Pulsar at right ascension 11 h 33 m), was found to emit a radio pulse once

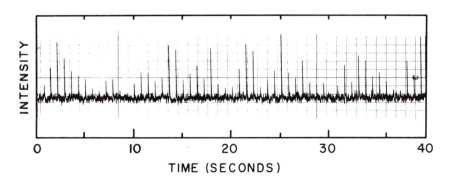

**FIGURE 11.1.** A sequence of radio pulses observed at a wavelength of 75 cm from PSR 0329+54, pulse period 0.715 s. These data were displayed on a chart paper, and the pulses were strong enough to emerge easily from the receiver noise fluctuation, indicated by the wiggly line extending the full length of the display. (R. N. Manchester.)

every 1.33730110168 seconds. The radio signals were so regular that at first the radio astronomers took seriously the possibility that they had detected messages from extraterrestrial intelligence, or "little green men," as they jokingly referred to the first four mystery sources.

The pulsating radio sources turned out to be anything but ET. They are related to extraordinary objects called neutron stars which spin incredibly rapidly and emit radio signals in beams which scan the heavens just as a lighthouse sends its light over the ocean.

## The Properties of Pulsars

The objects giving rise to the radio waves from the pulsating radio sources have never been seen, although X-rays from material falling onto them have been observed by satellite-borne X-ray telescopes. Knowledge of neutron stars is indirectly inferred from the study of the radio pulses. Also, two of 375 pulsars so far detected lie within supernova remnants and a couple are also seen to emit light pulses.

Pulsars transmit with extraordinary regularity, with most pulsar rates (periods) falling somewhere between ten times per second and once every three seconds. In a few cases the pulse rate has been observed to undergo a slight acceleration, known as a glitch, which is thought to relate to seismic activity which momentarily changes the shape of the neutron star and hence its spin rate. Pulsars have been observed across the radio spectrum from 10 to 15,000 MHz.

Today pulsars are named according to their location on the sky; for example, the pulsar PSR 1937+21 is located near right ascension 19h 37m, declination +21°.

# The Pulse Timing

Some of the recently discovered pulsars beat so regularly that they are as good as the best international time standards which are set by atomic clocks. The regularity of pulsar PSR 1937+21 has even exceeded our ability to measure it. Whatever causes the radio pulse is probably related to an object spinning with incredible steadiness. There is no other way to maintain such regularity in an astronomical context. Each pulsar has its characteristic pulse shape, which refers to the way the radio intensity varies during the pulse. These pulse shapes are the "signature" for each pulsar. From the duration of the pulse, compared to the time between pulses, it is found that the typical pulsar beam is between 10° and 20° wide.

All pulsars are slowing down almost imperceptibly, yet measurably. This is a natural consequence of aging through the loss of energy by radiation. Some of the really old pulsars even skip beats for minutes at a time. Pulse rates can be measured to an accuracy of a millionth of a microsecond ($10^{-12}$ s), and when observations are collected over several years a change in this period can be measured to within an accuracy of $10^{-18}$ s/s.

Many pulsars are observed to move across the sky by a very small amount from year to year. This is found from very accurate observations of pulse arrival times. The moving pulsars all appear headed away from the Milky Way at a few hundred km/s, and extrapolating back in time suggests that they were born about a million or so years ago, in the disk of the Milky Way.

Because of the very short periods and their relative faintness, pulsars are very difficult to detect. The first few were found because they were relatively strong radio sources and their periods of a few seconds allowed them to be discovered by the radio telescope system designed to observe radio source scintillations. However, a typical modern radio telescope, making routine observations, would not detect pulsars by accident, because the receivers are not adjusted to be sensitive to rapidly changing radio source intensities. They have what engineers call a "time constant," which averages signals over several seconds to minutes. Pulsar detection equipment attached to the world's largest radio telescopes is now very sophisticated and involves extensive computer-aided searching of the radio data in order to recognize the pulsar.

# Where Are the Pulsars?

The first step in figuring out what a pulsar is requires finding its location. Most pulsars are found fairly close to the Milky Way, so they must be part of our galaxy. Their distances are derived by several means. Interstellar hydrogen gas absorbs pulsar radio signals at 21 cm, and in the case of many pulsars, hydrogen line absorption measurements allow a distance to be derived. A more general way to estimate pulsar distances is related to a phenomenon known as *dispersion*. The pulsar signals are modified as they pass through interstellar space, in particular when they pass through regions of ionized hydrogen between the stars. Interstellar ionized hydrogen—the thermal electrons which coexist with other components in space—causes radio waves to be slightly slowed down compared to the speed of radio waves in a vacuum. The change in speed is different at different frequencies. A pulse produced at the pulsar will arrive at earth at slightly different times depending on the frequency. This dispersion is measured by observing the arrival times of the same pulse at slightly different frequencies. The amount of dispersion depends on the total number of electrons which affect the radio signal on its path to earth. Astronomers know that the average interstellar electron density is about $0.03 \text{ cm}^{-3}$, and so the distance to a pulsar can be found.

Most pulsars are located between 25,000 and 35,000 light-years from the galactic center, and up to 3000 light-years above or below the galactic disk. A pulsar has also been discovered in a nearby galaxy, the Large Magellanic Cloud.

The majority of pulsar beams do not happen to flash in our direction, and taking this into account suggests there may be 500,000 pulsars in the galaxy. They should therefore be born at the rate of about one every 10 years, which is odd, because supernovae, the likely source of pulsars, appear to be born at the rate of about one every 50–150 years.

## Formation of Neutron Stars

In 1969 a pulsar was discovered inside the Crab nebula (Figure 8.2), the remains of the supernova of 1054 A.D. This pulsar, PSR 0532+21, has also been detected optically and through X-rays and gamma rays. It flashes at a rate of about 33 pulses/s, and its existence is confirmation of the theory that pulsars are created in exploding stars.

Whatever the nature of the pulsar itself, it has to be spinning extremely fast. Normal stars would shatter long before they could spin as fast as pulsars. The only form of matter which is capable of accounting for the pulsar behavior is a neutron star—a star consisting entirely of neutrons.

Creation of a neutron star involves the catastrophic collapse of the core of a fairly normal star of at least four solar masses whose outer layers explode in a supernova. The explosion is the consequence of a sequence of events triggered when the core of the star runs out of fuel. At that stage the internal fire which kept the star going is extinguished, but until then it was the internally generated heat that balanced the inward pull of gravity. When the fire dies the core cools, gravity suddenly dominates, and the core collapses. In the collapse the fundamental particles of matter, protons and electrons, are driven so close together that they fuse and become neutrons and a solid ball of these particles is produced at the center of the star.

In some cases this collapse continues with such violence that the neutrons are forced even closer together and, in turn, swallow each other in their own gravitational pull. This is how black holes are formed. The existence of a black hole can be recognized when it is in an orbit about a companion star. The black hole may draw matter out of that star much like a vacuum cleaner sucks dust from some distance away. The gas will plummet toward the black hole and stream into an ever-decreasing orbit about it to form a (by now well-known to the reader) accretion disk. The gas in the disk heats up and radiates intense X-rays (observed from earth) before disappearing into the black hole itself.

Returning to the·machinations in the supernova event, the core has collapsed to form a neutron star (or sometimes a black hole). Layers of gas above the collapsed core suddenly find that there is nothing to hold them up against gravity. Momentarily they hang suspended and then crash downward, smashing onto the neutron mass and rebounding in a violent and fiery explosion. We may see the flash from earth.

The newly born neutron star will be spinning extremely rapidly as a natural consequence of its having contracted so much. This is due to the conservation of angular momentum, also displayed by a spinning ice skater who begins a spin with arms outstretched and then spins faster and faster as he draws his arms in. This action changes the effective radius of his spinning body, determined by how far his arms are outstretched. A diver doing somersaults uses the same principle by tucking her body in at the start of the somersaults, causing her to become a smaller object which rotates faster. When she stretches out just before entering the water, the somersaults are slowed to a near stop. In the case of the spinning neutron star, shrunk to some small size, its gravitational pull remains sufficient to hold the star together against the disruptive force of rotation.

## Neutron Star Properties

The neutron star is a truly bizarre object. It has a typical mass of about 1.5 solar masses, a radius of about 10 km, and an outer crust consisting of a km-thick solid, crystal-like structure, or lattice, made of the nuclei of atoms stripped of all their electrons. Isolated electrons form a "sea" amongst these nuclei.

Below this layer is a 4-km-thick inner crust which consists almost entirely of a solid lattice of pure neutrons, immersed in a sea of electrons and neutrons. Deeper down, below 11 km, there exist mostly neutrons, together with a few protons and electrons which surround an inner solid core consisting entirely of neutrons.

## How Pulses Are Generated

Pulsars are a wonderful playground for physicists. The explanation for pulsar phenomena involves lots of basic, as well as high-level, physics. (The rest of this section, which contains more detail than has been presented so far, is not essential to understanding anything that follows in this book.)

From the observation of polarization of the radio pulses it is inferred that the pulsars have magnetic field strengths of between a billion and a trillion gauss. (The earth's field is about 0.1 gauss and the sun's normal field is 1 gauss.) Why pulsar fields are this high is an open question. It is believed that the fields may be generated by more complex mechanisms than the simple field amplification expected to occur in a collapsing mass, as was originally believed.

The radio-emission mechanism for pulsars is related to the synchrotron process, with slight variations to be mentioned. The loss of energy associated with the emission of radio waves causes pulsars to slow down with age. This effect is so dramatic that the energy loss of the Crab nebula pulsar, inferred from its slowing down, is found to equal the amount of energy required to cause the surrounding supernova remnant to shine. The Crab pulsar, PSR 0521+21, is the engine that feeds the supernova remnant known as the Taurus A radio source.

To account for the emission of periodic radio waves, the pulsar's magnetic field must be tilted from the spin axis, much as the earth's field is tilted away from its rotation axis (Figure 11.2). (The earth's magnetic north pole is not at its geographic north pole either, but about 12° away from it.) The pulsar's magnetic field may be tilted as much as 90°. Every time the magnetic pole near which the radio-emitting region is located happens to point toward the earth we pick up a pulse of radio waves.

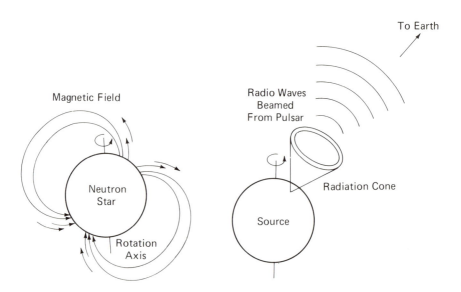

**FIGURE 11.2.** A schematic diagram of how a pulsar emits radiation. The magnetic field of the pulsar is tilted with respect to its rotation axis, and as the neutron star spins, the radiation beaming outward from the region near the magnetic pole of the neutron star sweeps by the earth once every rotation of the star.

The detailed radiation mechanism is not completely understood, but we will summarize some of the main points. A voltage or electric field is set up near the polar region of the pulsar due to its rapid rotation and strong magnetic field. One can imagine an analogy with a dynamo. Currents flow and particles (electrons) are accelerated outward along the pulsar's magnetic field lines. They produce gamma rays which, in turn, interact to produce pairs of electrons and positrons. This process continues and generates a cascade of particles which stream upward at close to the speed of light.

A few of these pairs, generated about 100 m, above the surface, send a particle back to the surface where they heat it and cause it to emit ions. This is called thermionic emission, and is very similar to the process that occurs inside vacuum tubes, which used to be part of all radio and TV sets. The ions react with X-rays produced at the hot spot to create more particle pairs which causes more particles to flow up along the field lines.

The energetic cascade of electrons and positrons forms clumps of particles, probably because of instabilities produced as the particles stream past each other. The clumps are forced to move along curved paths because of the overwhelming strength of the curved magnetic field close to the neutron star (Figure 11.2). This curvature causes the particles to radiate radio waves (just as particles spiraling about interstellar magnetic field lines do) from a region about 1000 km above the surface. The fact that the particles are clumped together greatly intensifies the strength of the radio waves, in a manner similar to that in a maser.

The glitches in pulsar periods are a mystery which continues to draw a lot of attention. It used to be believed that they were due to "starquakes," but one of several newer theories suggests that sometimes extra-large clouds of particles shower down onto the pulsar surface and the resulting heating causes changes in friction between the surface and the outer crust, which in turn causes a slight glitch in the way the pulsar spins.

## The Binary Pulsar—Nature's Fabulous Space Lab

During 1974 a major pulsar search was launched at Arecibo observatory. Joe Taylor and Russell Hulse devised an elegant technique which allowed them to discriminate against interference and quickly recognize a pulsar. They found 40 new candidates. One of these, in the constellation Aquila, was labeled PSR 1913+16, and it turned out to be very peculiar, even for a pulsar. The pulses occurred on an average of one per 0.05903000 seconds, but this rate was not constant, unlike all the other pulsars observed before. Its period first showed a 7-h 45-min cyclical change. PSR 1913+16 appeared to be a binary pulsar, a neutron star in orbit about another object. Pulse rate changes were produced by the Doppler effect (see the Appendix), which caused the arrival time of pulses to speed up or slow down (by 16.94 pulses/s with respect to the average) as the pulsar moved either toward or away from the earth during its orbit.

At least half the stars in the galaxy are locked in pairs known as *binaries;* therefore, a binary pulsar should not have been a surprise. However, the nature of this binary was extraordinary. Careful timing observations enabled the variations in the pulse arrival time to be interpreted with sufficient accuracy to allow the precise orbit, and subsequently masses, to be estimated. It was discovered that PSR 1913+16 consisted of two objects, each of about 1.4 solar masses, traveling about at hundreds of kilometers per second in orbits so close that the distance between them ranged from 1.1 to 4.8 times the radius of the sun (which is about 650,000 kilometers). The maximum diameter of the pulsar's orbit was only a million kilometers.

The binary pulsar provided a fabulous additional bonus. It is a perfect clock in orbit about a massive object, the ideal laboratory for testing Einstein's general theory of relativity. In 1915 Einstein had developed an elegant way to describe gravity and its effects and had explained an observation made during the previous century—that Mercury's orbit about the sun shows an anomaly not accounted for by other theories. Mercury's point of closest approach to the sun, known as its perihelion, moves slowly around the sun at a rate of 43 arcseconds per century. This is called the "precession" of the perihelion. Einstein's theory explained this phenomenon, and now the discovery of the binary pulsar provided a further test. The pulsar is in an elliptical orbit about another object and their point of closest approach (known as the periastron) should also precess. Changes in the pulse arrival times (due to the Doppler shift) should show tiny variations as the pulsar's orbit itself slowly swings around in space. The effect was measured to be 4°/year, far greater than for Mercury's movement about the sun, yet precisely as predicted by relativity theory.

The binary pulsar, however, turned out to offer an even more exciting prospect for the radio astronomers, an opportunity unique in the history of the science. They recognized a novel way to test one of Einstein's most important predictions, that objects accelerating in a strong gravitational field should emit a form of radiation called *gravitational waves*. Just as radio waves are produced by accelerating electrons, so gravity waves should be produced by accelerating matter. In the case of the binary pulsar, the conditions for radiating gravitational waves appeared to be perfect. Two massive objects were constantly accelerating within each other's gravitational influence. Einstein had stated that he believed gravitational waves would never be detected on earth because they are far too feeble to produce any measurable effects. Despite his caution, however, several laboratories have, with a notorious lack of success, attempted to detect gravitational waves. Now the radio astronomers realized they could search for the effect on the pulsar orbit as a *consequence* of the radiation of gravitational waves. They would not directly search for the waves, but see what happened to the binary orbit as the system lost energy in the form of gravitational radiation. The energy loss should be manifested as a very small change in the orbital period. This is the consequence of conservation of angular momentum, discussed before. As the system loses energy, its orbit shrinks; the pulsar will move a little faster through space, and hence the time taken to complete one orbit decreases.

Six years later, after extensive monitoring of the radio pulses from the invisible object in Aquila, Taylor and his coworkers reported that the pulsar orbital period was slowing down by $6.7 \times 10^{-8}$ s/yr, equivalent to a shrinkage of 3.1 mm/orbit or 3.5 m/yr. This was just the amount that should result from the radiation of gravitational waves. This remarkable measurement, confirming a prediction of a theory proposed 66 years earlier, has proven to be one of the most exciting bonuses produced by radio astronomy research.

But why are these two objects, the pulsar and its invisible companion, so close together in space? The other object is likely to be a neutron star, perhaps a pulsar, but its beam of radio waves does not happed to sweep past the earth. The two objects could not have been so close when they were normal stars. The explanation runs something as follows. Once they were two normal, albeit quite massive, stars, members of a binary system. The more massive one evolved quickly, consumed its fuel, and died in a violent supernova explosion. The neutron star stayed in orbit about the other star, which, in turn, reached old age and began to expand to form a red giant. The neutron star then became enveloped in the star's atmosphere, where it experienced frictional drag, slowed down, and slowly spiraled deeper into the giant star. The neutron star would not suffer undue hardship at this point, but would produce severe reactions in the giant star. In due course the star exploded and produced the second neutron core. Today the two neutron stars are in close orbit in the binary pulsar PSR 1913+16, nature's most remarkable laboratory in space.

Only five of the 375 or so known pulsars have been discovered to be members of binary systems. This is odd, because more than half the stars in the galaxy are members of binaries. Why aren't there more binary pulsars? What happened that caused most pulsars to be lone voyagers through space? Perhaps when the more massive star became a supernova it lost its companion in the process due to the sudden change in gravitational field produced by the formation of the neutron star and the enormous mass loss in the explosion. The pulsar and its former companion then separated to suddenly wander through space on their own. It appears to be rare for the neutron star to stay wedded to its companion after the falling-out induced by the explosion.

## The Millisecond Pulsar

The discovery of the millisecond pulsar is another tale involving persistence and following clues which could easily have been ignored. This story again begins with the scintillation of small-diameter radio sources. In addition to the effect described before, the apparent angular size of a radio source is made larger when the radio waves pass through clouds of electrons in interstellar space. These clouds produce what is known as interstellar scintillation, again most noticeable at low frequencies. Because of this effect, no very distant radio source, located close to the plane of the Milky Way, would show scintillation because interstellar scattering causes blurring, so that a pointlike radio source becomes a disklike image, which should not scintillate.

The clue that led to the discovery of the millisecond pulsar, which flashes at a rate of close to a thousand times per second, was a mysterious entry in a catalog of radio sources which indicated that a scintillating radio source, called 4C 21.53W, was located close to the galactic plane. According to the theory of interstellar scattering this should not be possible. Early pulsar searches showed no pulsar at the position of 4C 21.53W, so the reason for its scintillation continued to be a mystery. Perhaps the source represented a new class of object, which some wanted to call "scintars."

During subsequent research it was discovered that due to a rare error in the original survey, two radio sources were masquerading as one. This still did not explain why the source was apparently scintillating, but it did focus attention on discovering what the radio source looked like at high resolution—see Figure 11.3. These observations revealed that a tiny source appeared to be located next to a larger one. Furthermore, continued observations of the source did not always reveal the mystery scintillation effect. It turned out that the pulsar which did exist there was not readily discovered because its pulses were affected by interstellar scintillation. This caused the pulses to remain hidden for minutes at a time. Since no one expected pulses at the rate at which this neutron star was transmitting, the discovery was made even more difficult. The confusion was sorted out by the persistent work of Donald Backer at the University of California, Berkeley, with a team of collaborators in the U.S. and Europe. They opened their minds to the possibility that an extremely rapid pulsar was involved. That led to the discovery of the so-called millisecond pulsar, PSR 1937+21, which flashes at the rate of once every 0.0015578064488724 seconds.

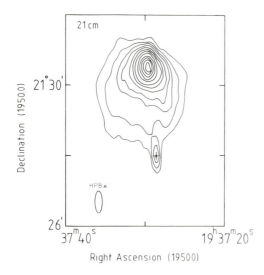

**FIGURE 11.3.** The location of the millisecond pulsar at the edge of an unrelated background radio source (4C 21.53W) as observed at 20-cm wavelength with the Westerbork Synthesis Radio Telescope in the Netherlands. The small ellipse labeled HPBW indicates the antenna resolution capability, about 13 by 37 arcseconds. (D. C. Backer. Reprinted by permission from *Nature*, Vol. 300, p. 615. Copyright © 1982 Macmillan Journals Limited.)

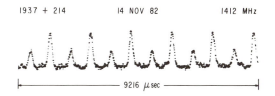

9216 μ sec

**FIGURE 11.4.** Radio pulses from the millisecond pulsar PSR 1937+21 observed at 20-cm wavelength with the Arecibo radio telescope. The length of this recording, 9216 microseconds, shows the extremely short time between pulses. (D. C. Backer. Reprinted by permission from *Nature,* Vol. 300, p. 615. Copyright © 1982 Macmillan Journals Limited.)

Its pulses are shown in Figure 11.4. This pulsar is located 16,000 light-years away in the constellation Vulpecula.

The pulse frequency is about 642 Hz, or E above high C on the piano. Recordings of PSR 1937+21 made at the 1000-foot-diameter Arecibo radio telescope have converted the radio wave into a sound wave, which allows the high-pitched humming sound of the pulsar to be heard quite clearly. Its staggering pulse rate is faster than any other pulsar, and according to conventional pulsar theory a neutron star spinning this fast should be very young—but no supernova remnant is located at its position.

Young pulsars are expected to run down quickly, yet the period of PSR 1937+21 is constant to one part in $10^{19}$, which implies that it is very old. Why this apparent contradiction? Because the millisecond pulsar is believed to be a recycled pulsar! It must once have been a normal pulsar, a member of a binary system. As its companion aged and swelled in size toward the end of its life, the neutron star may have gobbled up its companion. The two may literally have blended and in the process the old pulsar's spin rate was greatly speeded up, because in the process of absorbing that matter it has to spin faster in order to conserve angular momentum.

## What Pulse Timing Tells Us

The millisecond pulsar provides astronomers with the most accurate clock in the universe. It is not subject to the complex variations observed in the binary pulsar timing experiments. This clock is so accurate that the millisecond pulsar allows a whole host of other phenomena to be explored. Accurate timing of the pulse arrival over several years has allowed the position of the pulsar to be pinpointed to a staggering degree of accuracy. Its right ascension is 19h 37m 28.74600s and declination 21° 28′ 0.1406″ (1950). The pulsar is found to be moving across the sky by about a milliarcsecond per year.

Millisecond pulsars also provide a test for the theory of relativity. For example, the delay in pulse arrival times due to the gravitational effect of the sun on the radio wave as it travels through space is potentially detectable, as is the changing gravitational field in the vicinity of the earth as it moves in a slightly elliptical orbit about the sun. Time dilation effects (the way a clock slows down the faster it travels), due to changes in the earth's orbital velocity, have already been measured and support Einstein's prediction of this effect. Through continued observations it will become possible to discover whether gravitational waves are sweeping into the earth, causing it to "shudder" ever so slightly, or whether they are striking the distant pulsars as they move through space. Such waves may cause an otherwise imperceptible wobble in the earth's, or the pulsar's, motion, and a disturbance as small as that expected from this "buffeting" may be revealed by long-term observation of the binary pulsar.

PSR 1937+21 produces such accurate pulses that Don Backer was able to state that there is no "rock around the clock," which means that no mass of any significance, not even asteroid-sized, appears to be in orbit about it.

A second millisecond pulsar, PSR 1953+29, has been discovered and is found to be a member of a binary. It has a period of 6.133 ms and is in a 120-day orbit about its companion, a 0.3-solar-mass object, probably a white dwarf star.

No one knows whether more millisecond pulsars will be discovered, but when a half dozen are available for continued observation, they will provide astronomers with a remarkable tool. Since the regular pulses allow pulsar positions to be accurately measured, they can, in turn, be used to accurately locate the radio telescopes with respect to the direction of the pulsars. Using several millisecond pulsars, this could be done with extraordinary accuracy. This would give information on earth motions, and continental drift, and would allow the creation of a basic system for establishing time standards which could not be equaled by any terrestrial clock.

## Are We Still Open to the Unexpected?

Pulsars would not have been discovered if human intuition and persistence had not be directly involved in the process. The increased use of super-sophisticated computer technology in making astronomical observations now seems to remove the human factor to the point where serendipitous discovery is well-nigh impossible. A computer can only do precisely what we want it to do, which means that we have to know what we are looking for before we search! This implies that important discoveries such as Jocelyn Bell's discovery of "little bits of scruff" might not be possible in the 1980s. Our technological sophistication is now so great that we may no longer be as open to the unexpected.

# 12

# The Radio Sun and Planets

## War Secrets

On the morning of February 12, 1942, the German warships Scharnhorst and Gneisenau passed undetected through the English Channel on a voyage from Brest in France to Kiel in Germany. The reason they sailed unmolested by British warplanes was that the British radar was being jammed by some source of radio interference. J. S. Hey, a physicist who had learned something about newly invented radar at the outbreak of World War II, was assigned the task of investigating the jamming. The suspicion was that the Germans had come up with a device that could blind the British radar.

A few weeks later widespread jamming again occurred and the military responded by going on extreme alert, yet no hostile action followed. Hey discovered that jamming had occurred only in the daytime, when the radar antennas had been pointed in the general direction of the sun. A check with the Royal Observatory at Greenwich revealed that at the same time a large sunspot group was visible on the solar surface. Like Jansky before him, Hey found that extraterrestrial radio signals were responsible for unwanted "interference" and so he wrote a secret memo reporting that the jamming seemed to be produced by radio signals from the sun. Thus, out of the desperate situation of World War II, the seeds of solar radio astronomy were sown.

## The Plasma Sun

The sun is a churning mass of hot ionized gas with magnetic fields threading their way through every pore and core, driven by energies boiling out from the interior where the fusion of hydrogen into helium (at a temperature of $15 \times 10^6$ K) liberates the nuclear energy that keeps the caldron boiling. Heat bubbles toward the surface, which maintains a temperature of about 6000 K. A gas at this temperature radiates primarily light (and some heat and ultraviolet radiation—as we know from personal experience) by the thermal emission process.

A cloud of hot ionized gas, called a *plasma,* can support a large number of wave motions within its volume. When ionized particles move in harmony they begin to radiate energy, and the generation of radio waves by the solar plasma oscillations is of fundamental importance to the understanding of solar radio waves. The level of this understanding has reached such awesome proportions that one prominent solar physicist has suggested that the theory of solar radio emission is now so well developed that most astronomers can no longer understand it! This need not concern us, because we will only touch upon the most important phenomena and leave the details to the experts.

The visible surface of the sun, known as the *photosphere,* is mottled by light and dark patterns where plasma actively surges up and down—cooling as it rises, heating as it falls. Occasionally small regions become unusually active. Magnetic fields tangle and knot and trigger more upheavals. The fields tear, twist, and turn, and bits and pieces intermingle and reconnect to form new patterns of force. The reconnection of magnetic fields is usually accompanied by the sudden release of vast amounts of energy, energy which was originally held in the fields and is then converted into the explosive ejection of particles into space. These explosions are observed as bright *flares* of light on the solar surface, often near cool sunspots, where magnetic field activity is particularly intense. The plasma around these magnetic field explosions is set into oscillation and radio waves are generated, which travel outward to reach the orbit of the earth eight minutes later. (The sun–earth distance is 8 light-minutes.) Figure 12.1 is a 20-cm radiograph of the solar disk, showing several radio sources over its surface. When sunspot activity is very great, and solar flares repeatedly burst out over the surface, intense radio noise is produced, so intense that human life can be affected.

During active spells, solar magnetic fields coil and uncoil, heave and churn and arch upwards. These arches are called *prominences* when seen at the edge of the sun. They rear up like uncoiling snakes, and great clouds of plasma are offered a way to escape the boiling heat below. Tentacles of magnetic field break and release their grip and clouds of particles stream out into space, triggering oscillations in the surrounding plasma as they rise into the solar atmosphere (called the *corona*).

The moving clouds successively trigger radio emission higher and higher in the corona, and occasionally clouds of ejected plasma reach the earth and impinge on its magnetic field. The earth's field acts as a shield, an invisible force field which protects us from solar particle storms. The traveling plasma clouds slide past the planet, leaving spaceship earth untouched by the harmful effects which would result if the plasma clouds were to crash unimpeded into our atmosphere. These high-energy particles can destroy ozone (the molecule $O_3$) that exists high in the atmosphere and protects us from direct solar ultraviolet radiation. Ozone absorbs ultraviolet radiation, which is fortunate for us because large doses of this radiation are fatal to terrestrial life forms.

After particularly violent solar storms, particles can penetrate the protective field, especially in the region of the earth's magnetic tail, which stretches out

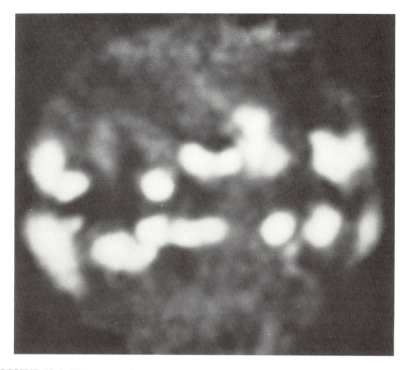

**FIGURE 12.1.** The sun as observed at 20-cm wavelength with the VLA on September 26, 1981, near the peak of the sunpot cycle. The bright portions of the diagram indicate regions radiating the greatest radio energy. (G. A. Dulk. Reproduced, with permission, from *Annual Review of Astronomy and Astrophysics,* Vol. 23. Copyright © 1985 by Annual Reviews Inc.)

beyond our planet like the wake behind a boat. The tail is swept there by the perpetual wind of particles blowing out of the sun. Following a solar storm, these particles (mostly electrons) may worm their way into the earth's magnetic tail, where they promptly rush helter-skelter along the magnetic field toward the polar regions of our planet. These electron streams crash violently into the highest regions of the terrestrial atmosphere where they collide with, and ionize, atoms of oxygen and nitrogen. These gases then vibrate with energy so that they produce the magnificent fiery displays known as *aurorae.*

## Solar Radio Emission

The study of solar radio waves was launched in earnest in the postwar years, when many physicists the world over salvaged surplus radar equipment whose antennas and receivers were ideal for studying the sun. Today the radio sun has been observed from millimeter to kilometer wavelengths, and modern solar radio astronomy is replete with extraordinary detail, so much that little point would be served in attempting to cover it all. Instead, we will mention only some of the highlights.

Of the three primary mechanisms by which the sun emits radio signals, two have already been mentioned in other contexts—synchrotron emission and thermal radiation from a hot plasma. A third mechanism, which has many variations, involves natural oscillations of the plasma itself. Radio emission can occur at the frequency of the plasma oscillations as well as multiples of this frequency.

## The Quiet Sun

Radio emission from the *quiet sun* is observed at times of sunspot minimum and comes from regions low in the corona. A *slowly varying component* may be observed which varies with the ponderous rotation of the sun, one cycle every 28 days. The variable intensity is partly related to the presence of visibly darker (cooler) regions known as coronal holes, which alternate with slightly warmer, more normal plasma over the surface of the sun. The quiet sun, by definition, is observed when there is little violent activity occurring.

The slowly varying component is also related to the presence of filaments of hotter gas which thread their way over the solar surface. In their immediate neighborhood, temperatures change from 6000 K in the filaments to 1,000,000 K in the surrounding corona, which has long been known to be extraordinarily hot.

The radio brightness of the sun (Figure 12.1), which is a measure of the temperature of the gas emitting the radio waves, ranges from 10,000 K at centimeter wavelengths—corresponding to regions low in the corona, close to the visible surface—to a few million degrees K at meter wavelengths high in the corona, and shows pronounced hot spots in the active regions (Figure 12.1).

## Solar Radio Bursts

On a regular basis, especially when many sunspots are present, regions on the surface of the sun may grow steadily hotter and brighter until a flare occurs. Near such *active regions,* whether or not accompanied by flare activity, bursts of radio waves may be generated.

Figure 12.2 shows a schematic diagram that summarizes the nature of a variety of *solar radio bursts,* sudden increases of radio energy over a range of frequencies and lasting for some time (from seconds to days). Figure 12.3 illustrates how some of these bursts are formed. Five different types of radio burst are most common.

Type III bursts are produced by plasma clouds streaming up into the corona and triggering oscillations as they go. At any instant, emission occurs over a narrow range of frequencies in a band which shifts to lower frequencies as the

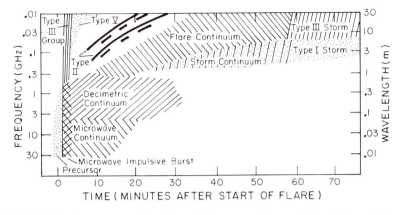

**FIGURE 12.2.** A summary of the time and frequency characteristics of the main forms of solar radio-burst emission (see text). In general, the way the frequency of the radiation decreases with time is a direct indicator of the motion of the source of radiation upward through the solar atmosphere. (G. A. Dulk. Reproduced, with permission, from *Annual Review of Astronomy and Astrophysics,* Vol. 23. Copyright © 1985 by Annual Reviews Inc.)

cloud moves rapidly (at $\frac{1}{3}$ the speed of light) into space. The motion of the rising clouds is studied by observing the way the frequency of the emitted solar radio waves shifts with time. The frequency of plasma oscillations depends on the electron density, which decreases with height in the corona. Therefore, the frequency of radio transmission from a cloud traveling up through the solar corona shifts to lower bands as the cloud rises. Sometimes astronomers can track ejected clouds as far as the earth, where satellites have probed them directly. Near the earth, plasma oscillations generate radio signals at a frequency of 20 kHz (wavelength 15,000 m).

Type II bursts move more slowly and are produced by clouds of plasma traveling away from the active region at about 900 km/s. Strong plasma turbulence is believed to create Type III bursts, and a shock wave then triggers surrounding plasma oscillations which generate Type II bursts (Figure 12.3).

Storm bursts (Type I—not shown in Figure 12.3) are brief spikes of intense radio emission, each lasting about a second, which accompany violent solar activity associated with sunspots or flares. Solar *noise storms* may last hours or days, with tens of thousands of individual Type I bursts transmitted during this time.

Bursts labeled "continuum" in Figure 12.2 radiate a broad level of activity over a wide frequency band. They show no indication of movement, and they are not spiky like Types II and III bursts. The continuum bursts are created by the synchrotron process, which generates radio waves over a wide frequency range. Another phenomenon, Type V continuum bursts, usually follows Type II bursts.

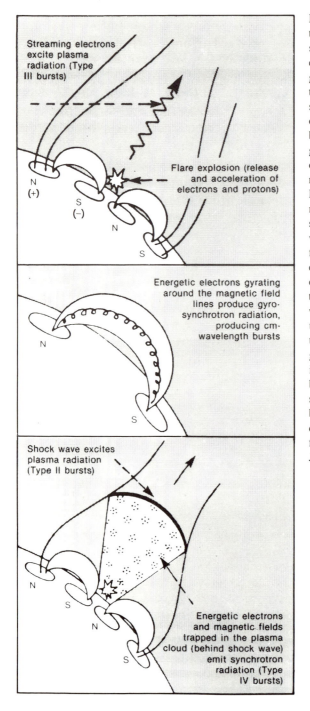

Streaming electrons
excite plasma
radiation (Type
III bursts)

Flare explosion (release
and acceleration of
electrons and protons)

N
(+)

S
(−)

N

S

Energetic electrons gyrating
around the magnetic field
lines produce gyro-
synchrotron radiation,
producing cm-
wavelength bursts

N

S

Shock wave excites
plasma radiation
(Type II bursts)

N

S

N

S

Energetic electrons
and magnetic fields
trapped in the plasma
cloud (behind shock wave)
emit synchrotron
radiation (Type
IV bursts)

**FIGURE 12.3.** Generation of solar radio emission. Top frame: A flare explosion occurs in the region of magnetic field interaction between two sunspots. Particles streaming outward trigger disturbances in the surrounding gas, which trigger plasma oscillations and generate radio waves. Center frame: Electrons gyrating about magnetic fields between sunspots generate cm-wavelength bursts. Lower frame: Following a flare, clouds of particles stream outward. Some of them are trapped behind the shock wave (the boundary between the expanding matter and the surrounding gas) and emit radio energy in the form of Type IV bursts. Particles in the shock generate Type II bursts. (M. Kundu. Reproduced, with permission, from *Sky and Telescope*. July, 1982 page 6.)

Decimetric continuum bursts, also known as Type IV (which follow Type II), are related to synchrotron emission from electrons trapped in magnetic fields around sunspots. A variety of wave motions may be set up in these magnetic fields regions (called magnetic tubes). When these fields bulge outward or expand or undergo some other catastrophic changes, Type IV bursts (Figure 12.3) may be produced. These also do not drift in frequency.

## Radio Signals from the Planets

All objects at reasonable "everyday" temperatures emit radio waves. This includes the moon (Figure 12.4), the earth, the planets, and your own body. The dark universe beyond the stars is at a temperature of 3 K (see Chapter

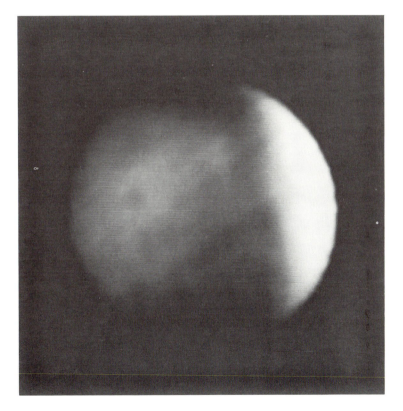

**FIGURE 12.4.** The moon as observed at about 3-mm wavelength with the 12-m radio telescope shown in Figure 10.3. The moon's phase was just past new. The brighter crescent is the part of the moon illuminated by the sun, where the higher temperature produces more thermal emission. The resolution of this radiograph is 30 arcseconds. (NRAO. Observer—C. J. Salter)

14). The earth, at a temperature of about 290 K (about 60° Fahrenheit), would appear as a thermal radio source to a distant radio astronomer. The fact that all objects at a finite temperature emit radio waves by the thermal process means that even if a radio telescope is pointed at the ground or at a distant clump of trees, it will pick up radio waves.

Radio astronomers expected thermal emission from the planets, and therefore our neighbors in space became some of the first research targets. Mercury, Mars, and the moon are known to be relatively normal sources of radio emission, their radio brightness being those expected from objects at their temperatures. (The temperatures of the planets are directly inferred from heat, or infrared, measurements.) Venus, however, produced a surprise. Its cloud tops are at a temperature of 230 K, but in 1956 the first radio observations of this planet showed that its temperature was 600 K (about 900° Fahrenheit). This discovery came as a considerable shock. The Venus cloud layers apparently act as a greenhouse, keeping the surface of the planet at 600 K, a temperature later confirmed by direct measurements from landing spacecraft.

## Jupiter's Radio Bursts

In 1955, as part of a 22-MHz (13.6-m wavelength) sky survey, two budding radio astronomers (B. F. Burke and K. L. Franklin, working at the Carnegie Institute in Washington, D.C.) made daily observations of the Crab nebula (Taurus A), which they used as a standard signal to calibrate their data. As they further developed the antennas and receivers of their equipment, they persisted in monitoring the Crab. When it came time to begin the systematic search for new radio sources, they had to make daily adjustments to the antenna system so that it would receive radio signals from directions a little further south.

The arbitrary decision to start their mapping program by pointing the telescope further south (rather than north) paved the way for their major discovery. Unknown to them, Jupiter was lurking up there and moving a little further south each day. Soon their data began to reveal unwanted "interference" (a theme familiar to us by now). The interference came through soon after the Crab nebula was observed. After a few days of this, they began to take the signals seriously and sought an explanation. A colleague, Howard Tatel, apparently jokingly, suggested it was Jupiter.

On that same evening, out in the field where the antennas were located, Burke noticed a bright object in the twilight sky and asked his partner what it might be. "Jupiter," came Franklin's answer, causing them to laugh at the odd coincidence in view of the remark by Tatel earlier in the day. Neither noticed that Jupiter was in Gemini, the constellation immediately adjacent to Taurus, the home of the Crab nebula.

The next day Franklin, perhaps in desperation, decided to explore the Jupiter connection more closely. To his complete surprise, he found that, indeed, Jupiter could be blamed for the "interference." Jupiter's radio signals turned out to be not steady emissions, such as might be produced by the thermal or synchrotron process, but intense bursts, not unlike those produced by the sun. This discovery, so utterly surprising, was one of the most unexpected in radio astronomy. By chance, the peak energy in the radio bursts is concentrated in the radio band around 20 MHz. If Burke and Franklin had been observing at 40 MHz or higher, or at another time of year, or if they had started to survey to the north of the Crab nebula, the radio bursts would not have been discovered until years afterwards.

The story has an ironic twist. Australian radio astronomers had been observing the radio sky at 19 MHz and years before had noticed a peculiar source of radio emission, but its cause had remained a mystery to them. Their antennas did not have sufficient resolution to pinpoint the source, and privately they believed that perhaps the swishing sounds they heard were terrestrial interference originating somewhere over Indonesia. With the announcement of Burke and Franklin's discovery, the Australian researchers looked back at their old records and found that Jupiter had produced the "interference" and that Jupiter was visible on records going back five years. Thus, within weeks of the discovery that Jupiter was a radio source, they had five years of data to work with (a situation seldom met with in science).

The low-frequency Jovian radio waves are known as its *decameter* radiation, since the wavelengths of the bursts range from 10 to hundreds of meters. Jupiter serves up such intense doses of radio emission at these frequencies that its bursts can be heard on a loudspeaker attached to even a modest radio telescope and receiver.

Up to this time Jupiter had been considered to be a dull, cold, giant ball of gas, yet now it was suddenly a focus of intense interest. By 1958 the first theories to explain Jupiter's decametric bursts had been proposed. They included discharges from lightning in Jovian thunderstorms, plasma waves generated by storms in its atmosphere, or oscillations in the ionized regions of the high atmosphere triggered by volcanic activity on the planet.

In 1964 it was discovered that one of Jupiter's large satellites, its moon Io, plays a role in determining when the burst radiation is beamed in our direction. It is now known that there are at least four regions which produce slightly more intense radio emission. None of these regions are physically fixed on Jupiter. The emission mechanism is complicated, barely understood, and apparently related to plasma wave phenomena. The radio-burst radiation is beamed along narrow angles, so that when the beams sweep by the earth, the signals are observed to vary in intensity. Whether we pick up bursts depends on where the earth is in Jupiter's sky and on Io's location with respect to the Jupiter–earth line. The presence of Io in the appropriate location appears to enhance the intensity of radio waves beamed in our direction.

However, back in the 1950s, Jupiter provided radio astronomers with another considerable shock.

## Jupiter's Radiation Belts

Jupiter, the largest planet in the solar system, is very cold because it is so far from the sun. Infrared studies of Jupiter had shown the temperature of its cloud tops to be 150 K (as compared to 220 K for the earth's cloud tops). Radio observations at 3-cm wavelength showed that Jupiter behaved like a thermal radio source at that temperature. Despite the discovery of the radio bursts, which were clearly being generated by some nonthermal process, Jupiter was expected to be a normal thermal emitter at other wavelengths. However, at 10 cm Jupiter's radio temperature was found to be 600 K, and at 70 cm it was 70,000 K and this radio emission was polarized. This incredible discovery meant that in addition to being an emitter of thermal radiation, Jupiter was also a nonthermal radio source and was behaving like distant radio sources in which cosmic-ray electrons spiral around magnetic fields to produce radio waves. No one had expected this.

The early observations of Jupiter's nonthermal radiation were carried out by Frank Drake and Hein Hvatum at the National Radio Astronomy Observatory. The NRAO had just been established and had only one radio telescope, the 85-foot Tatel Telescope, coincidentally named after the same man who later contributed to the discovery of the Jupiter bursts.

Jupiter is a strong nonthermal radio source in the wavelength range 3 m–10 cm; this is called Jupiter's *decimetric* radiation. The explanation for the short-wavelength radio emissions was forthcoming after a major discovery by artificial satellites designed to detect cosmic rays near the earth.

In 1958, Explorers 1 and 3 revealed that the terrestrial magnetic field thousands of miles up contains trapped high-energy particles (mostly electrons), which spiral around the field lines and move north and south, between the polar regions of the planet. They make up the so-called van Allen radiation belts, named after their discoverer. The region around a planet in which the magnetic field is important is known as the *magnetosphere,* and the van Allen belts are part of the magnetosphere. It was not long before it was realized that processes occurring in Jupiter's magnetosphere might be the source of its decimetric radio waves. If Jupiter had a strong magnetic field containing trapped particles, these particles could produce nonthermal radio signals. This was shown to be what was happening, and today radio emission from the Jovian radiation belts is directly observable with high-resolution radio telescopes. Figure 12.5 shows radiographs of Jupiter which confirm the theory that the radiation belts are the source of the radio waves. The belts reach from about 90,000 to 200,000 km above Jupiter's cloud surface. The radio waves are polarized, and because Jupiter's magnetic field is tilted by 10° with respect to its rotation axis, the

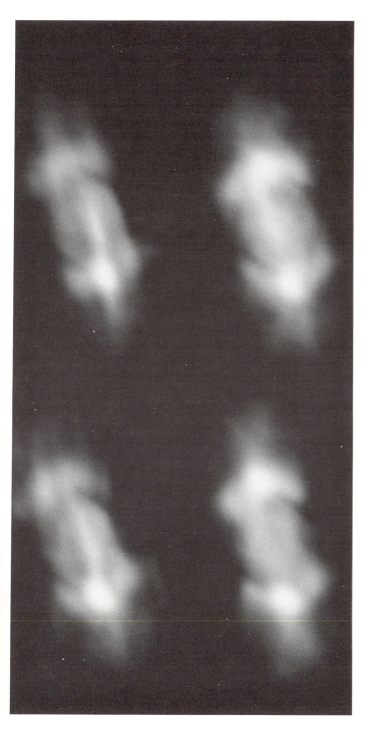

**FIGURE 12.5.** Several radiographs of Jupiter's 20-cm-wavelength emission, made at various times during its rotation period of 9 h 55 min. Thermal emission from the disk of the planet can be seen while strong (nonthermal) emission originates in two lobes on either side of the planet, produced by the edge-on view of the Jovian radiation belts. The radiographs were made as the planet rotated by 15°, 105°, and 120° with respect to the upper left image, moving counterclockwise from upper right. (NRAO. Observers—J. A. Roberts, G. L. Berge, and R. C. Bignell.)

intensity of the received radio signals varies with time and gives a very accurate measure of the planet's rotation period: 9 h 55 min 29.71 s.

## Jupiter's Kilometer Radiation

By the 1970s Jupiter was already a very well-studied radio source, but there was more to come with the launching of various earth-orbiting spacecraft equipped with radio antennas and receivers to study Jupiter's very-low-frequency radiation from above the earth's ionosphere (the electrically conducting layer high in the atmosphere through which low-frequency radio waves do not pass). In 1973–74 the Radio Astronomy Explorer I and IMP6 satellites discovered that Jupiter was emitting radiation at very long wavelengths, but in the earth's environment the radio spectrum was found to be very noisy because of electrical activity in the auroral regions near the earth's poles. This local noise prevented a clear view of Jupiter. It was necessary to get away from the earth in order to study Jupiter's kilometer radiation more carefully. Thus, in 1978–79, the Voyager spacecraft carried radio astronomy equipment to the planet for a closer look.

Kilometer-wavelength signals began to be picked up 300 million miles from Jupiter. The signals were found to reveal an extraordinarily complex picture of bursts shifting in frequency and time. A broadband burst source in the frequency range 10–100 kHz (wavelength 30–3 km) is strongly modulated by Jupiter's rotation period and showed that the source was confined to high latitudes on the daylight side of the planet. The origin of these bursts is still not known. A second source, a narrow-band one peaking at around 100 kHz (wavelength 3 km), occurs periodically and shows a cyclical change which is not the same as the planet's rotation rate. This may be because the plasma at the source drags behind the rotation of the planet itself. The emission appears or disappears for days at a time, and neither this phenomenon nor the emission mechanism is understood.

## The Planets as Radio Sources

Jupiter is the best-studied radio source in the sky, after the sun. The detailed nature of its wide variety of radio emissions is now known to be extremely complex. It has been observed over 24 octaves in the radio spectrum, from frequencies of 10 kHz to 300,000 MHz and, in summary, several things are very well known while many mysteries remain. It is known that thermal emission from Jupiter's atmosphere accounts for all the very-high-frequency radiation at wavelengths below about 7 cm. Synchrotron emission from trapped high-energy particles in the magnetosphere (Jupiter's radiation belts) dominates the radio spectrum from 4000 to 400 MHz (wavelength 7.5–75 cm). The theory of this radiation is well understood. Sporadic low-frequency emission below 400 MHz (longer than 75-cm wavelength) is due to plasma phenomena known as instabilities. The theory of these emissions is hardly understood at all, although it is known that any burst-type radiation originating at frequencies beyond 39.5 MHz would not be able to escape the plasma in which it is generated (where it is absorbed instead) and hence this frequency represents a real cutoff.

Saturn and the earth also generate radio emission similar to the Jupiter bursts. The peak of Saturn's burst radiation occurs at around 500 kHz and the earth's at around 60 Hz. Voyager observations of Saturn showed very-low-frequency bursts which have been explained in terms of possible lightninglike discharges in Saturn's rings.

To a radio astronomer at the outer reaches of the solar system, the earth would appear as the strongest radio source in the sky at 60 Hz. Those radio signals originate in the auroral regions, but the signals are not observable down here on the earth's surface because such radio waves cannot penetrate the electrically charged ionosphere in the upper reaches of our atmosphere. It is an odd coincidence that worldwide sources of commercial alternating current (AC) are distributed at 50 or 60 Hz, a frequency which happens to coincide with the earth's natural frequency of burst radiation.

## Radar Astronomy

Up to this point we have described the discoveries of radio astronomers, a passive breed who can do no more than receive radio signals which are naturally transmitted by distant objects in the universe. However, radar astronomers bounce signals from the moon, planets, large asteroids, and the sun and listen for the echoes.

Echo location is a wonderful technique used by several species in the animal kingdom to help them "see" their otherwise invisible surroundings. Dolphins and bats transmit sound pulses and receive echoes which allow them to sense obstacles and prey. *Homo sapiens* transmits pulses of radio energy and receives echoes from the planets in order to locate them. The remarkable human species uses the largest single-dish radio telescope on earth, the 1000-foot-diameter dish near Arecibo, Puerto Rico, to do the job (see Figure 17.1).

Ever since World War II, radar has been a powerful tool for locating and observing the motion of objects in the air, on land, and at sea. During the war, radar signals were already bounced off the moon. Today radar is used to map the surface of Venus, a study which certainly fits into our theme of an invisible universe revealed. The Venusian surface is otherwise completely hidden below a perpetual cloud cover.

Radar astronomers transmit pulsed radio waves at a particular frequency and oberve the reflected signal returned to their receivers. The delay time between transmission of the pulse and reception of the echo tells them how far away the reflecting object is. If the target is moving, the frequency of the echo is also shifted with respect to the frequency of the transmitted signal. This frequency shift is another manifestation of the Doppler effect. The Doppler shift of echoes from a moving object is utilized in a similar way in radar (traffic) speed traps.

The intensity of the radio energy reflected back from the surface of a planet depends on how smooth the surface is. If you shine a flashlight onto a bowling ball you will see a highlight from the surface nearest you, where most of the light is reflected. Because the ball is highly polished, the light which strikes it outside the region of the highlight bounces off in other directions. Now shine the flashlight onto a tennis ball with a rough surface. The tennis ball has many bits of fur which scatter light in all directions, and since this fur covers the entire surface, there is always some of it to reflect light in your direction.

This is what happens when radio signals are reflected off a planet. Some parts of the planet shine more brightly than others, depending on their texture or their orientation with respect to the radar beam. The roughest surfaces appear as the brightest ones in radar images. A steep mountain slope near the edge of the planet's disk will reflect more radio waves back to earth if it is angled so as to better mirror these signals directly toward us.

Radar astronomers have invented beautifully sophisticated ways of creating pulsed radio signals which allow the echoes to be interpreted in great detail. The transmissions are pulsed according to special codes and the computer searches for these patterns in the echoes. Transmission can last up to five minutes before the echoes return and are received for the following five minutes.

## Planetary Rotation

One of the first experiments in planetary radar, back in 1965, resulted in the discovery that Mercury rotates in an unexpected manner. A moving radar target, such as a planet, shifts the frequency of the reflected radar signal relative to the transmitted signal. However, a planet also rotates, which means that some part of its surface will be moving toward us and the adjacent side will be moving away from us. Due to the rotation, the reflected radio signals will be more than just Doppler shifted—they will be spread out in frequency, that is, Doppler *broadened*. Thus the effect of planetary rotation can be found in the planetary echoes.

Mercury radar observations showed that Mercury rotates once in 59 days and does not keep its same face to the sun, something which had been believed for centuries, in which case its rotation period would be the same as its revolution period around the sun, 88 days.

Venus radar observations showed that it rotates once in 243 days in a direction opposite to that characteristic of the other planets. It also presents alternate faces from one rotation to the next and always shows the same face to earth at the time of closest encounter.

FIGURE 12.6. A radar image of a portion of the optically invisible surface of Venus made with the Arecibo telescope (Figure 17.1) in 1975 and 1977. Within the empty band insufficient data were available to make a good image. The brightest areas are due to surfaces whose roughness are comparable with the 12-cm wavelength of the radar observations. The smallest details that can be recognized are about 10–20 km

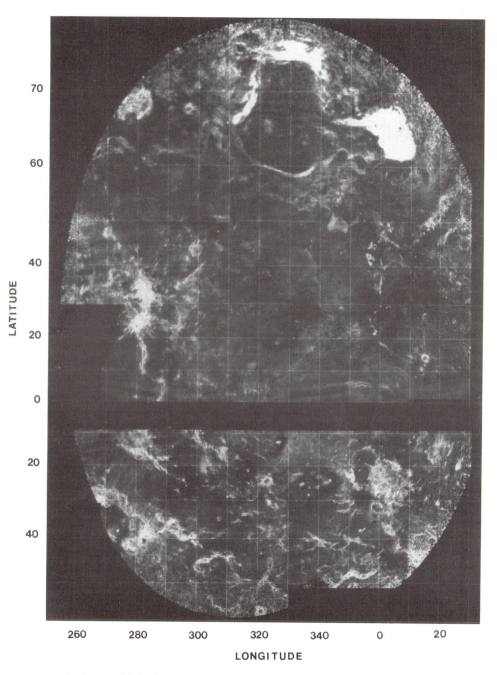

across (a degree of latitude on Venus equals about 106 kilometers.) The bright area at the top right is known as Maxwell Mons; the two patches at latitude 30° at the left-hand edge are called Rhea and Theia Mons, in what is known as the Beta region. Three craters can be recognized in the dark area around longitude 340° and latitude −25°. (D. B. Campbell, Arecibo Observatory.)

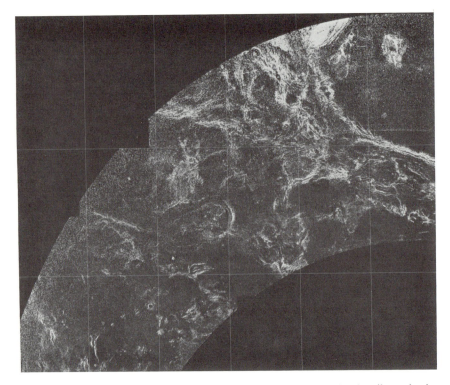

**FIGURE 12.7.** A swath of the Venusian surface observed with the Arecibo radar in 1983; see text. The left-hand image extends from Venusian longitude 270° to 330°, latitude 40° N to 70° N. The grid marks are every 10°. The right-hand image extends from longitude 330° to 25°, latitude 39° to 67° with latitudes lines indicated at 49° N and 59° N. The dark area in the nearly overlapping central region is Planum Lakshmi, an elevated area a few kilometers higher than the average surface level. This region can be recognized in the upper center of Figure 12.6. Several small round structures, either impact or volcanic craters, can be seen at various locations. (D. B. Campbell, Arecibo Observatory)

## The Face of the Goddess

For two weeks every 19 months, when Venus comes relatively close to the earth, radar astronomers at Arecibo observatory instruct engineers to turn on a jet engine which powers an electrical generator supplying their 400,000-watt radio transmitter. Then, as long as Venus is within view of the 1000-foot-radio telescope, they blast a 12.5-cm-wavelength radio signal at our sister planet. The echoes are collected by the same radio telescope and contain trivial amounts of power (only $10^{-20}$ watts), yet they carry enough information to allow the

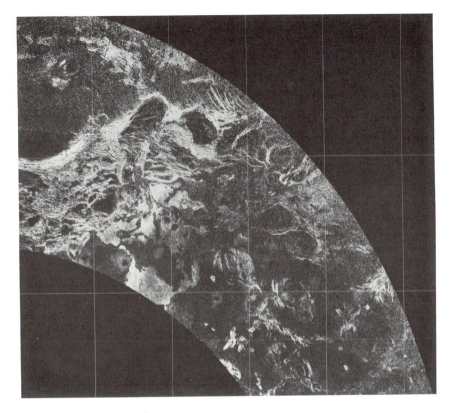

**FIGURE 12.7.** (*continued*)

surface of Venus to be mapped from earth with an accuracy that cannot be achieved by any other means (see Figure 12.6).

Venus is perpetually covered in a thick layer of carbon dioxide clouds and so its surface is hidden from us. The Pioneer Venus Radar Mapper has made radar maps of the Venusian surface to reveal details 100 km across, but surface details only a few kilometers across and a few hundred meters high can be mapped with the Arecibo telescope.

Venus radar studies revealed three major highlands on a planet otherwise extremely flat. Beta Regio, Aphrodite Terra, and Maxwell Mons are three large elevated regions on the surface of this planet as shown in the radar map made in 1975 and 1977, Figure 12.6. Maxwell Mons is 11 km high, higher than Mt. Everest.

Figure 12.7 is a map of a swath across the northern part of Venus, made during 1983. Several small craters can be most easily spotted at the left- and right-hand sides of these maps. The dark area at the center is known as Planum Lakshmi, an elevated region which can also be recognized in Figure 12.6 at the upper center of that larger map. To the right and below the dark area in

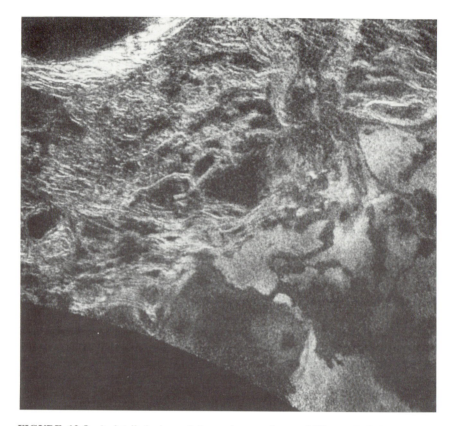

**FIGURE 12.8.** A detailed view of the region southeast of Planum Lakshmi (Figure 12.7), showing bright terrain which is generally very flat and appears bright because its surface has irregularities of the order of the 12-cm observing wavelength. The dark, wavy patterns are probably due to old lava flows. (D. B. Campbell, Arecibo Observatory.)

Figure 12.7 a very bright region signifies that the surface is very reflective to radio waves of 12-cm wavelength, which gives information about the rockiness of the terrain on this scale. This region is a large, gently sloping plain. The flowlike patterns, shown in more detail in Figure 12.8, are believed to indicate old lava courses.

## Other Radar Targets

Mars has also been scanned with radar beams, but it is the radar echoes from more elusive objects in the solar system that have added the most new data to the astronomer's archives. The larger moons of Jupiter (Callisto, Ganymede, and Europa) reflect a tremendous amount of radar energy, far more than the planets or our moon. Europa reflects 65% of the energy back and acts so much like a mirror that to the radar astronomer it is indistinguishable from a rough metal sphere. The precise nature of the reflected signals gives astronomers a lot of information about surface textures, and the echo properties of the Jovian satellites can best be explained if the surface consists of extensive fields of broken ice.

Radar echoes from Saturn's rings show that they are made of chunks of ice about 10 cm and more across. Radar echoes have also been obtained from some of the larger asteroids whose surfaces, shapes, and rotation are thus measured. Two earth-crossing asteroids—that is, asteroids whose orbits cross that of earth—one known as 2100 Ra-Shalom and the other as 1685 Toro, have diameters of about 3 km, an awesome size if the earth were ever to encounter such objects in a direct collision. Radar echoes from many comets, including Halley, have been obtained and such experiments continue. Comet nuclei are shown to be about 1 km across.

## The Solar System and Beyond

The early history of active radio astronomy research shows that the sun and the planets were a great focus of activity. More recently, however, the sun and Jupiter have been particularly intensely studied. As was mentioned in connection with solar radio waves, the theory is so well understood that only the specialists can understand it. Earthbound radio observations of the sun and planets now continue at a reduced pace as compared to the explosion of radio astronomy research in other areas. Instead, the plasma physicists and theoretically inclined spend a lot of time trying to explain the subtle physics of the sun or Jupiter's magnetospheric phenomena. However, the sun is only one of hundreds of billions of stars in the galaxy, and when suitable radio telescopes were built in the 1960s and 1970s, it did not take radio astronomers long to ask which other stars might be significant radio sources.

# 13

## The Galactic Superstars

### The Curious Object SS433

We have journeyed through the invisible radio universe from the distant quasars and radio galaxies, via the galactic center and interstellar space, to our solar system. Some remarkable radio phenomena with a common underlying theme have been revealed; for example, the universe witnesses a wide variety of highly energetic explosive events, usually associated, it is theorized, with very compact objects such as black holes, which eject remarkable jets of material ablaze with radio energy. But what about the $2.5 \times 10^{11}$ stars in our Milky Way, stars which otherwise appear to be relatively normal? A closer look reveals some extraordinary surprises, not the least of which was the discovery of SS433, first believed to be an innocuous little star, one of countless others in the direction of the constellation Aquila, a little star which has since proven to be the most bizarre object in the galaxy.

When SS433 was first noticed in the 1960s it appeared to be no more than a faint red star, except that it showed hydrogen emission lines. These are spectral lines generated by hot hydrogen at the surface of the star, an interesting phenomenon to two astronomers, C. Bruce Stephenson and Nicholas Sanduleak. This star was the 433d entry in their catalog of such objects.

In 1976 X-rays were discovered to be coming from the direction of SS433 and then, in 1977, radio waves were observed from the same position. This sounded an alarm in the minds of many astronomers. A star emitting unusual amounts of both X-rays and radio waves deserved a closer look. What was revealed stunned the astronomical community—SS433 is a small-scale version of the phenomenon powering radio galaxies and quasars!

SS433 is located 18,000 light-years away inside an old supernova remnant first observed in the late 1950s and known as W50, whose radio portrait is shown in Figure 13.1. The remnant is believed to be about 40,000 years old and its size has swollen to 200 light-years across, enveloping thousands of stars in the process.

In 1978 routine studies of the spectral lines emitted by this peculiar star were begun in order to see if these would give a clue as to why the object

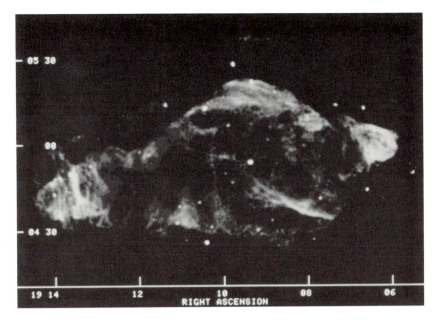

**FIGURE 13.1.** Radiograph of the supernova remnant W50 surrounding the strange object SS433, whose radio emission produces the small bright spot in the center of the region (right ascension 19h 09m 20s and declination 4° 55′). The radio source is elliptical in shape and about 2° × 1° in extent with brighter "ears" at each end. SS433 lies at the center of the nebula and its radio jets (see Figure 13.3) align with the major axis of the nebula. It has been suggested that W50 is a supernova remnant that has been ejected in the east–west directions by the force of the jets from SS433. (NRAO. Observers—S. A. Baum and R. Elston.)

emitted such strong radio waves and X-rays. The first detailed observations were so startling that the astronomers involved thought that something had gone wrong with their equipment! This single "star" showed three sets of spectral lines, quite unprecedented in astronomy. One set was apparently perfectly normal, showing a small Doppler shift of 70 km/s, expected for the star's direction and distance in the galaxy (see Chapter 9). This indicated that the star was partaking of relatively normal motion with respect to all other stars. However, the other two sets of spectral lines were bizarre. One set indicated an extraordinarily high redshift, indicating motion away from the earth at 50,000 km/s. If this were to be interpreted as a typical redshift (as is observed in distant galaxies and quasars), the object would have to be over 1,000,000,000 light-years away, hardly a star in our galaxy! The other set of lines showed a blueshift (motion toward us) of 30,000 km/s. The star appeared to be both moving away and toward us, at some sizable fraction of the speed of light, even though it simultaneously appeared to be moving normally! But this very

peculiar beast had more shocks in store for astronomers, who began to flock
to their telescopes by the dozens to observe this cosmic wonder.

Repeated observations revealed that the spectral lines were not constant in
time. They showed an amazingly regular change in their Doppler shift (which
meant a systematic change in velocity), as is shown in Figure 13.2. One set
of lines varied between a redshift of 50,000 to 0 km/s while the other set
varied between a blueshift of 30,000 km/s and a redshift of 20,000 km/s. The
cycle repeated every 164 days.

Figure 13.2 also reveals that the average Doppler shift for the spectral lines
is about 12,000 km/s. Why isn't it close to zero, the velocity of the star itself?
After all, the third set of spectral lines showed only a small Doppler shift of
70 km/s due to the star partaking of normal galactic rotation.

These observations were interpreted in terms of a by now familiar phenomenon.
The stunning stellar fireworks display generated by SS433 could be explained
if it were ejecting two jets of luminous material, just like quasars and radio
galaxies. Astronomers were suddenly confronting the fact that a quasarlike
object existed in our galactic backyard. Apart from the fact that it was clearly
much smaller than a radio galaxy, SS433 looked just like an extragalactic radio

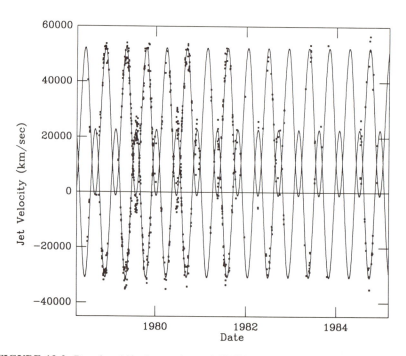

**FIGURE 13.2.** Doppler shift observations of SS433 taken over a six-year period. Two
sets of spectral lines show enormous systematic velocity variations, one set ranging
between 0 and 50,000 km/s, the other between −30,000 and 20,000 km/s (see text).
(B. Margon, University of Washington)

source with two jets reaching into space, and it was no doubt also powered by some remarkable central engine. Furthermore, because the gas in these jets was moving so fast, another important discovery was made. According to relativity theory, if the jets are produced by material moving at a significant fraction of the speed of light across the sky, and not just either away or toward us, a second phenomenon comes into play. This is the so-called *transverse Doppler effect,* which is a shift in the spectral lines due to motion transverse to our view of the moving object. There is no everyday analog of this phenomenon, which only becomes important when the object travels at a significant fraction of the speed of light. The variation in the spectral lines due to the jets could be explained if material in the jets were streaming at about 75,000 km/s, an incredible 25% of the speed of light; this would produce a transverse Doppler shift of 12,000 km/s, the average observed for the spectral lines.

## An Explanation for SS433

Two jets appear to be blasting away from the central compact object, which may be a neutron star, but is far more probably a black hole. A relatively normal star is in orbit about it, a conclusion founded on further observations which showed periodic 13-day changes in the spectral line velocities from SS433. The source of the jets is therefore a member of a binary star system. But to what, then, is the 164-day period due? The cause cannot be related to a neutron star, because they spin at the rate of once per second. A black hole would spin even faster (if the spin of such an object could be detected). Of course, no normal star, which rotates so slowly, could expel jets with such violence. This leaves only one culprit, the accretion disk around a black hole.

## The Black Hole and Its Accretion Disk

The SS433 system most likely consists of a normal star in orbit about a black hole of about 4 solar masses. Due to the close proximity of the nearby star, the black hole lures material from the surface of the normal star and draws it outward, toward the black hole. As the infalling material gains speed, it begins to gather into the familiar accretion disk, in which particles undergo violent collisions with each other. This disk is the halfway house between former stardom and the black hole's world of nothingness. However, there is a wonderful twist to this story.

If too much material rushes into the accretion disk, a condition known as *supercitical accretion* is reached and then things become very interesting indeed. An enormous increase in particle collisions suddenly heats the gas to the point where it contains so much energy that it explodes, driving material outward again, to escape the clutches of the black hole. But this material cannot blast through the surrounding material in the disk; it can only go up the central hole of the doughnut-shaped accretion disk. So away we go again; two jets blasting outward, this time demonstrably at a quarter of the speed of light, a very chaotic state of affairs.

The two jets tear into space, gathering up material as they go (through the process of entrainment, which we confronted earlier in Chapter 5), and gather up more energy. The jets also expand sideways at about 2000 km/s and fan out slowly as they rush into the surrounding supernova remnant. The jets themselves appear to be something like long, miniature cylindrical supernova remnants! The hot material in the jet also emits X-rays, as observed by X-ray astronomy satellites in 1976.

According to this picture, SS433 should show two nice straight jets pointed away from the central source, in which the velocity of material streaming outward would remain constant with time. But they don't look like that. The radio observations of SS433 (Figure 13.3) show that the jets are not straight, but look like corkscrews whose twisting motion can be followed from day to day; and the reason for this cosmic corkscrew is related to the 164-day period in the jet velocities.

## Precession of the Accretion Disk

The accretion disk appears to be wobbling about the black hole just as we have found to be the case in distant quasars. The precession period for the disk around SS433 is 164 days (unlike the hundreds of millions of years which is the case for the supergigantic black holes and accretion disks at the nuclei of quasars).

Why does this disk precess? The binary star companion of SS433 is feeding the voracious black hole an excessive diet of gas, mostly hydrogen, and at the same time pulling on the accretion disk. That should be enough to cause slow precession, but there is more to it. The star itself is not round! Due to the proximity of the black hole it is distorted, and therefore the gravitational influence from the rotating, distorted star is very nonuniform. Note that the accretion disk surrounding this 4-solar-mass black hole is only solar-system-size and the black hole a few kilometers in diameter. (All these numbers, and details of the picture described here, come from a tremendous amount of research into the spectral line shifts and a comparison of all the observations made at optical, radio, and X-ray wavelengths, which are then compared against theoretical calculations.)

The result of the tug-of-war between the ugly star and the accretion disk is precession, which is like the wobble of a top set spinning on a table. However, the entire disk does not really move as a solid object, because the gas is passing through the disk so rapidly that today's accretion disk is almost a new one compared to yesterday's. Individual particles follow complex paths into the region of the black hole, and the net result is most easily pictured in terms of a precessing disk.

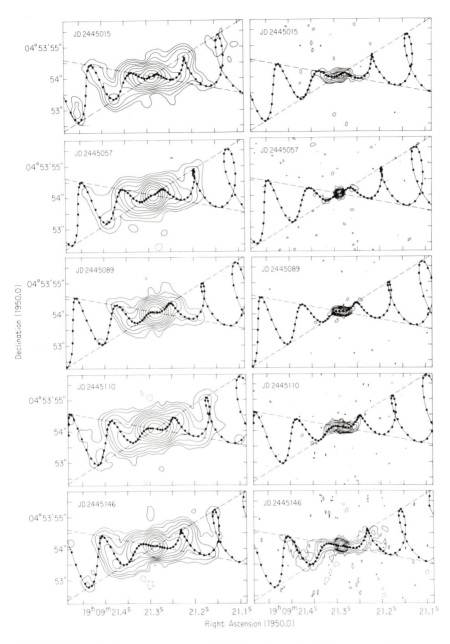

**FIGURE 13.3.** Radio contour maps of SS433 made at the VLA on several days over a four-month period. The day number, according to astronomical convention, is indicated at the upper left of each frame. The twisted lines indicate the theoretically predicted paths of material streaming out along the two jets at a quarter of the speed of light. The jets precess about the central star. The agreement between the theory and observations is regarded as extremely good. (NRAO. Observer—R. M. Hjellming.)

Now the particles in the two jets are ejected straight out into space, but because the orientation of the disk changes with time, the direction of ejection also changes; and so, even as individual particles head straight outward, they create the corkscrew-like pattern of radio emission. Their trajectories may be likened to water streaming out of a rotating garden sprinkler. Each water drop heads straight out, but as the sprinkler spins, it produces an apparent spiral of ejected liquid.

Seen from earth, the velocity of material in the SS433 jets cycles through a range of values determined by the geometry of the twisted jets with respect to our point of view. Sometimes a jet would point more directly toward us and days later it would be tilted away from us. This happens with the same precession period as that of the accretion disk, once in 164 days.

To summarize, the strange spectral lines from SS433 are produced by two jets of incandescent gas driven out of an accretion disk surrounding a black hole which is in orbit about a star which supplies the fuel! The jets are driven explosively outward by the energy created in supercritical accretion, which occurs when too much gas is made available for the black hole to swallow in one gulp. The jets, in turn, are propelled to the outskirts of the surrounding supernova remnant, which they keep fed with energy that makes the remnant shine.

The radio observations have led to a determination of the distance to SS433. Since the velocity of material along the jets is known and the movement across the sky can be seen in the radio maps, the distance to the object has been determined—18,000 light-years. The light we now see from this remarkable object has been traveling since *homo sapiens* dwelt in caves back in the last ice age, when humans were oblivious of the remarkable cosmic wonders that exist beyond the stars overhead.

The study of this apparently innocuous little star called SS433, has given us our first clear insight into the physical processes occurring near black holes. SS433 is still being thoroughly studied, and already the picture to account for its behavior is about as complete as any in astronomy, which is all too often spiced with mysteries which cannot be solved with present-day observations and always seem to require bigger telescope. SS433 is also a wonderful manifestation of the phenomenon occurring in radio galaxies and quasars, but here it is on a tiny scale, very close to home. Its discovery has reinforced the notion that jets and precessing accretion disks are enormously widespread in the universe.

## Another SS433?

Changes in the light and radio emission from SS433 suggest that this object will not last forever, but will die away in only a few thousand years. Similar objects are likely to be rare, although at least one other may have been observed less directly. This second example is the X-ray source known as Cyg-X3.

Following two strong outbursts of X-rays in October 1982 and again in 1983, Cyg-X3 showed two moving regions of radio emission in its immediate vicinity. Outward motions appeared to be at $\frac{1}{3}$ the speed of light. Unfortunately, Cyg-X3 lies in a heavily obscured region of the galaxy and so is hidden behind dust clouds and cannot be seen optically. Its spinning jets, if they are present, will be forever hidden from us.

## Radio Stars

The pathological object SS433 is certainly the strangest astronomical phenomenon ever observed in our Milky Way, but what about the other 250,000,000,000 stars? Each star is expected to emit radio signals by the thermal emission processes for no other reason than that the star's surface is hot, at a temperature somewhere between a few thousand and a few tens of thousands of degrees Kelvin. Very hot stellar atmospheres, up to several millions of degrees, are also common. The radio emission from the majority of stars like the sun is not detectable on earth because those stars are too far away, and hence their signals are too faint, to be detected. If the sun were placed at a distance of four or five light-years, the distance of the next nearest stars, its radio signals would barely register a flicker with the world's largest radio telescopes. Compared to radio galaxies, quasars, supernovae, and HII regions, the sun and other "normal" stars are radio quiet and virtually invisible. Nevertheless, there are several categories of stars, known as *radio stars,* which do emit radio waves, and each category does so for different reasons. Some of them even involve jets which have been directly observed.

The names of the various classes of radio stars are as colorful as the variety of phenomena involved. Irregular and infrequent radio blasts are emitted by RS Canis Venatici (RS CVn) or Algol-type binaries, M supergiants, UV Ceti-type flare stars, AM Herculis stars, symbiotic stars, novae, and VV Cephei stars. Each generates peculiarly intense radio signals.

Many radio stars generate thermal radio emission in strong winds of gas blowing out of the star or in ejected spherical envelopes of material expanding away from the star's surface.

The most commonly observed radio stars are members of binary systems known as RS CVn binaries, which have orbital periods of 1–30 days. Usually one star is like the sun and the other slightly older. These stars interact directly with each other, with magnetic fields from one star penetrating the other and particle clouds surging between them. Their radio emission is intense and highly variable; about 30 have been detected.

Some stars—for example, the variable star UV Ceti—appear to have atmospheres as hot as 10,000,000 K. This is a single star, with no binary companions, and was one of the first stars to be seen to exhibit giant flares, which generate intense nonthermal radio emission. These flares are far more violent than any flare on the sun. During a flare the light from UV Ceti can increase in brightness within minutes, before dying away again. Such a flare appears to be similar to a solar flare, and it became a great challenge to pick up its radio signals, even in the days before the technology really allowed a successful experiment. The problem was partly due to technological limitations, but also to the incredible faintness of the radio burst, which would have been all but indistinguishable from ground-based electrical interference and so would have made the observations highly suspect. Only when the large radio telescopes of the late 1970s and the 1980s began operations did flare star observation become reliable.

Flare stars are common amongst T Tauri stars (see Chapter 10), which vary in brightness and occur in intimate association with interstellar molecular clouds. These stars are believed to be very young, between 100,000 and 1,000,000 years old. They may be surrounded by accretion disks and seem to be blasting out matter in the form of minor jets which drive outward and push up against surrounding interstellar matter, where diffuse nebulosity may be produced (as was discussed in Chapter 10).

It is now believed that all stars may pass through this flaring phase in their early childhood. The optical flaring is easy to see, and even small telescopes can detect this activity.

There is a worldwide network of dedicated amateur astronomers who perform a valuable service by constantly monitoring a great many variable stars, including flare stars, a function that cannot be assigned to large telescopes whose time is divided amongst so many different research projects. The variable-star observers keep the rest of the astronomical community appraised of any specially dramatic events in the heavens. Their discoveries then prompt the professional astronomers to turn their larger telescopes to the relevant stars.

## Novae

The nova* experience is quite unlike that of a supernova. The latter is the complete destruction of a star, while the former is the mere shrugging off of an outer layer of gas in a relatively minor convulsion, but one which would destroy our planet should the sun ever go nova, an unlikely event according to current theoretical knowledge of stellar evolution. Afterwards the star may resume its normal existence or may repeat the process, in which case it is known as a recurrent nova. Radio emission from novae is produced in the ejected circumstellar envelope, and two expanding nova shells (HR Del and FH Ser), which give valuable data on the physics of such shells, have been constantly monitored for several years.

Early in 1985 (January 26) the recurrent nova RS Ophiuchus (RS Oph) was observed to undergo an optical tantrum, and over the next few weeks radio astronomers at Jodrell Bank in England and at the VLA in New Mexico saw

---

* See also Chapter 8, p. 87.

the radio emission rising and then slowly falling. A month later the ejected shell started to run out of energy and RS faded back to its quiet state.

## Other Superstars

Another marvelous phenomenon has been discovered in binary star systems. One star heats the particle wind blowing out from the other star. For example, the star Alpha Scorpii (Alpha Sco) has two radio components associated with two stars in the binary. A pointlike radio source is situated at the location of the smaller member, and a small nebula is observed around the relatively more massive companion star. The nebula is produced by the ionization of the wind blowing from the smaller star as it moves past the larger one. The larger star does not produce a significant outflow of gas in its own stellar wind, but does generate a lot of ionizing ultraviolet radiation. The smaller star produces a strong wind, but very little ultraviolet. Through teamwork they create a fascinating double radio source. This is an example of a symbiotic star. Figure 13.4 demonstrates a well-studied case, with two radio maps of the star R Aquarii (R Aqr) and an optical image in the center frame. There is bright emission from the star and a probable jet, visible as a patch of radio emission protruding from the star. The whole area around this object is further embedded in luminosity not seen in these images. It appears that mass exchange is occurring between two stars, one a 387-day variable, located within the large patch of emission, and the other a hot dwarf star very close to it. The radio emission from the jet at the upper left first appeared sometime between 1970 and 1977. The radio source did not significantly change its appearance between 1982 and 1984, which suggests that whatever was ejected has been dramatically slowed down by surrounding matter.

Mass transfer between stars—between a star and a neutron star or between a star and a black hole—appears to be common in our galaxy. In all examples of mass transfer, the gas is heated up to millions of degrees and emits X-rays as well as radio and light. Sometimes the mass being transferred is heated up en route to the other star, or as it streams past that star.

## A Star with a Jet

Early in 1985, the star CH Cygni (Figure 13.5) was observed to be ejecting a radio jet. No black hole is involved in its production. CH Cyg was known to consist of a red giant star in orbit about a white dwarf, and thus the jet is produced by two stars playing around. CH Cyg has exhibited repeated nova-like outbursts over many decades, but in 1984 its radio brightness again suddenly increased while it was being studied as part of an ongoing program designed to monitor about 100 symbiotic stars. (Weak radio emission had already been discovered from about a quarter of them.) CH Cyg emitted a burst of radio

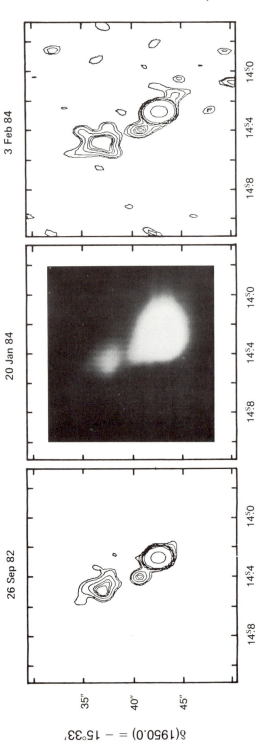

**FIGURE 13.4.** VLA maps of the radio emission from the radio jets associated with the R Aquarii, a star 1000 light-years away. The central frame shows an optical image of this object. Two larger sources are believed to be the jets emerging from the central, weaker source, which is due to thermal radiation from the star itself. (NRAO. Observers— J. M. Hollis, M. Kafatos, A. G. Michalitsianos, and H. A. McAlister.)

**FIGURE 13.5.** Radiograph of the radio jet associated with the variable star CH Cygni. Vertical size 36''. (NRAO. Observers—A. R. Taylor and E. R. Seaquist.)

waves and months later, when the star was mapped with the VLA, it first showed a double structure—75 days later it was a triple source, shown in Figure 13.5. These observations indicate that a jet of matter was ejected from one of the stars.

The astronomers studying this object believe that an accretion disk is again involved and that this star underwent an explosion in the disk which blew out material with considerably less energy than is exhibited in SS433, yet the same sort of phenomenon may be involved.

Clear evidence for a jet emerging from a star has, therefore, been added to the Gallery of Cosmic Jets, which now includes quasars, radio galaxies, the galactic center, T Tauri stars in molecular clouds, SS433, and binary stars.

In theory, all stars should be observable with large enough and sensitive enough radio telescopes. However, up to now radio emission from "normal" stars has not been observed. Such objects, if they emit as the sun does, are below the sensitivity limits of even the world's largest radio telescopes. Instead, every time radio emission from a star is detected the star turns out to be a peculiar object. No doubt, in the years to come, wonderful new species will be added to this astronomical zoo.

Now we are ready to move back out again, out to the universe beyond the quasars, to a time when the universe began, to see what the radio telescopes reveal about those remote reaches of space and time.

# Part V

## The Universe and Life

# 14

## Beyond the Quasars—Radio Cosmology

### A Cosmic Perspective

Our journey through the invisible universe has taken us from the realm of the radio galaxy and quasar toward the galactic center. We have traveled past supernovae and HII regions, seen giant molecular clouds, and listened to pulsars. All of these are awesome phenomena, yet the most extraordinary phenomenon may be that we are capable of perceiving these things and asking cosmological questions.

Cosmology deals with the origin and structure of the universe. Who has not read or heard some astronomer's claim that the next generation of telescope will allow us to see to the edge of space and solve some basic cosmological question? Such claims are perhaps based on hope rather than reality, because the human mind has hardly begun to probe the true nature of the universe. All we can do is stand before our discoveries and allow a feeling of awe to envelop us as we struggle to comprehend the universe on its largest scale. The universe is vast and we are but tiny specks upon one of her countless planets. Yet we believe that we can find answers to cosmological ponderings, and in this chapter we will consider how radio astronomy has some bearing on these issues.

The universe appears to be about 15–20 billion years old, a conclusion based on a knowledge of the ages of stars and the fact that the universe is expanding. Reversal of the present expansion rate would lead to all matter being concentrated at one point that long ago.

Consider that 12,000 years ago humans still lived in caves, sheltered from a dying ice age. Viewed from a cosmic perspective, 12,000 years is nothing—less than one-millionth of the age of the universe. Development of the scientific method for exploring "reality" was only initiated about 400 years ago. Seen from the cosmic perspective, our species has hardly stirred in its cradle. If the 15,000,000,000 years since the Big Bang were compressed into a single "cosmic year," a century of our time would be equivalent to one-fifth of a cosmic second.

Modern astronomy, armed with amazing telescopes capable of operating at all wavelengths within the electromagnetic spectrum, has only just begun to glimpse the contents of our universe. It was as recently as 1920 that the existence of distant galaxies first became widely accepted. Our view of an expanding universe and the Big Bang, which triggered the expansion, are fledglings on the scene of human thought. Only since the 1950s, after the advent of radio astronomy and the subsequent development of X-ray, ultraviolet, infrared, and gamma-ray astronomy, has our understanding of the contents of the universe truly begun to flourish.

At least two positions regarding the remarkable phenomenon of human consciousness and our awareness of our universe could be adopted. On the one hand, you might lean toward the notion that since we are so clever and ingenious, we are close to discovering all we would ever want to know about the origin and evolution of the universe and all in it, including life. In that case the next few millenia will, at best, be spent in cleaning up loose ends—and then what?

On the other hand, you might adopt a more conservative position and consider that the universe is far more complex, mysterious, and awesome than our human brains can comprehend. In which case, why bother? Yet this point of view need not stop us from searching further, of continuing our explorations, because curiosity is clearly a deep instinct. Curiosity and exploratory behavior may even drive evolution. The only problem is that today the satisfaction of human curiosity seems to cost so much!

In either case, though, it might behoove us to consider that our brains, when pitted against an extraordinary universe, might take longer than a fraction of a cosmic second to come up with all the answers. But can the typical scientist afford to adopt this stance when it comes time to justify a budget request for another dramatic experiment?

This chapter will not be presented in terms of "gee whiz, look how we are about to solve the cosmological questions." The real "gee whiz" is that we know very little about cosmology. Humans have barely begun to explore the astronomical universe.

## Radio Astronomy and Cosmology

Astronomers are convinced that the universe is expanding. This conviction is based on the observation of a universal redshift phenomenon. Distant galaxies, whose distances are derived in a variety of ways, are moving away from us, and the galaxies farthest away are receding most rapidly. This expansion of the universe is quite literally that. It is not just that the galaxies are moving apart in space, but also that space is stretching out between them.

Radio astronomical observations can contribute to our knowledge of the universe on the largest scale. For example, if the universe is expanding, then it might be possible to discover whether radio sources were different in the early universe as compared with our present epoch. This expectation held high hopes in the 1960s, when quasars and radio galaxies were first discovered. Many radio sources are beyond the range of the world's largest optical telescopes.

About 40% of radio sources are as yet unidentified, which means that there are no optically visible objects associated with them. Perhaps radio astronomers are in a better position to say something about the universe at the greatest distances. This is the same as suggesting that radio astronomers might be able to probe further back in time. Thus, in the 1960s, radio cosmology research began with great enthusiasm.

When radio sources still appeared as single (or at most double) points of radio emission, it was assumed that they were intrinsically the same sort of objects, and hence the distribution of sources over the sky could be studied. However, the radiographs and contour maps of distant radio sources shown in this book illustrate the point that radio sources now show a tremendous diversity. These objects do not represent a simple, uniform sample, a population which would be ideally suited for radio astronomers to do the best cosmology. On the contrary, such a sample may not exist anywhere in the universe. The distribution of radio sources in space does not appear to lend itself to any broadbrush interpretations, so it may not contribute to the cosmological study. In fact, there is no unambiguous evidence to suggest that any aspect of radio sources depends on their distance, that is, on the epoch in which the sources were born. The only thing we know is that quasars, in general, are farther away than radio galaxies. Quasars existed when the universe was slightly younger and smaller and matter was packed more densely, and so the quasars had more fuel available from the cannibalization of nearby objects. Thus they might shine a little more brightly.

Another radio test for cosmology, which was extremely popular in the 1960s, was the so-called "log $N$–log $S$" relationship. The cosmologically inclined radio astronomers delighted in counting the number ($N$) of radio sources to discover how this number depended on the brightness, or flux density ($S$) of the sources. The hypothesis was that as one counted fainter and fainter sources, one was going deeper and deeper into space, and therefore one should find more and more sources. The way the number of sources increases with decreasing flux levels (the variation of number $N$ with flux $S$) should give cosmological information. For example, if the universe was expanding and had a definite starting point, there might be fewer sources as one looked closer to the beginning. In a universe with no beginning and in which space was not expanding, the number of sources per unit volume would be constant, and would produce a different log $N$–log $S$ relationship.

The original source-count studies provoked a great deal of controversy, and today the situation seems to be at an impasse. The majority opinion now seems to be that the "log $N$–log $S$" data are not conclusive, probably because the inherent variation in radio source properties creates so much confusion that no unambiguous conclusions can be reached.

Figure 14.1 shows a radio map of a small area of sky containing dozens of faint sources. This is typical of what would be seen almost anywhere in the heavens where no strong foreground source dominates the picture. Counting radio sources requires that many maps like this be made to see how many

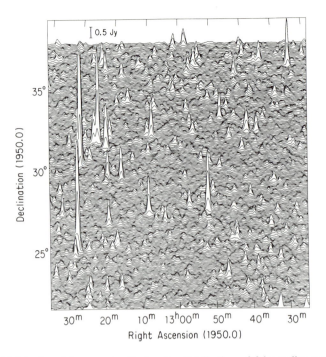

**FIGURE 14.1.** A striking way to display the distribution of faint radio sources on the sky, as mapped by the 300-foot telescope of the National Radio Astronomy Observatory (NRAO). Each horizontal trace indicates the output of the radio telescope as it was systematically moved across the sky. The sequence of horizontal "scans" maps the area of sky, and the presence of radio sources is indicated by the vertical deflection of the signal from the zero level in any individual trace. These data can also be converted into radiograph form. (NRAO. Observer—J. J. Condon.)

faint sources exist for any given number of brighter ones. A radiograph version of similar data is shown in Figure 14.2

## The Microwave Background

There is one radio observation, however, which contributes decisive input into the cosmological discussions. In 1963, observations by Arno Penzias and Robert Wilson, at Bell Laboratories, the birthplace of radio astronomy, revealed that our world is bathed in a 3 K glow of microwaves. We are all stewing in this mellow glow which represents a faint memory of conditions dating back to the creation of the universe. The discovery of the microwave background is an interesting case history in scientific progress, because when the radio signal was first detected it was suspected of being due to some spurious effects in the radio receiving system itself.

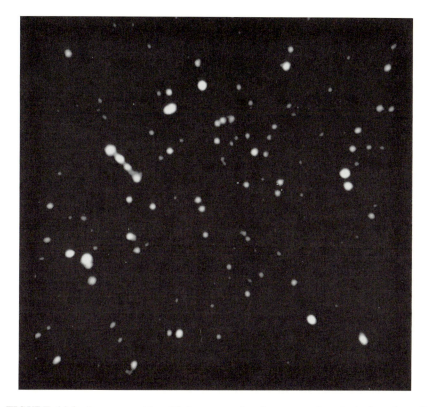

**FIGURE 14.2.** A very sensitive VLA survey of radio sources in a typical section of sky. Each of the starlike objects is really a very distant radio source, probably a radio galaxy or a quasar. These sources are so faint that no work has been done yet to try and identify them. The intensity of each dot in the image is related to the deflection that would be produced by each radio source in a diagram such as Figure 14.1. Vertical size is 20'. (NRAO. Observer—J. J. Condon.)

The story begins before World War II, when the physicist George Gamow, interested in explaining how the various elements came to be formed, hypothesized that the universe started as a fireball in which the elements were cooked up. Much later, Robert Dicke and his colleagues at Princeton University were planning observations to search for evidence to indicate whether we live in an oscillating universe which may have gone through a hot phase. Penzias and Wilson, meanwhile, were doing engineering work with a very sensitive radio antenna at Bell Labs.

Arno Penzias had been drawn to Bell Labs in 1961 with the promise that if he could get a horn antenna (a satellite communication antenna used in the Telstar project to send TV signals across the Atlantic) to operate at really high efficiency he could later use the antenna for radio astronomy. Robert Wilson joined him soon after, and together they were able to eliminate many sources of noise which lessened the efficiency of the antenna. When they made a list of all sources of noise, which included the sky, the ground, cables, and the radio receiver connected to the antenna, they were left with some residual noise, equivalent to that which would be produced by a blackbody (that is, a perfect absorber) radiating at about 3 K. In order to eliminate this, they explored further and took the horn antenna apart. Two pigeons, contributing about half a degree of the mystery signal, were nesting inside. The pigeons were shipped out of town, but they soon returned—they were homing pigeons! After the birds were barred from entering the horn antenna again, the astronomers returned to their measurements because they still had nearly 3° of mystery noise to contend with.

Penzias and Wilson believed that space was empty and shouldn't be radiating at 3 K. They did not know that Dicke was searching for evidence that the universe might be hot, nor did they know that Gamow had wanted a hot universe to cook elements. Meanwhile, two Soviet scientists, A. G. Dovoshkevich and I. D. Novikov, suggested that if certain theories of element formation were correct, the early universe should be radiating a detectable radio signal. This work was unknown to any of the above. So there they were! Penzias and Wilson found a signal, but didn't know what it was, and Dicke and his team were preparing to search for such a signal and didn't know it had been found!

Penzias and Wilson wanted to publish their discovery in an obscure place, since they didn't know what their signal was and suspected that it might be due to an error of measurement. For a while they were torn between not publishing the data or relegating the discovery to relative obscurity in a technical, nonastronomical publication about antennas.

Meanwhile, another astronomer, James Peebles, independently predicted that a radio signal of 10 K should be expected from the Big Bang. As a result of conversations between astronomers and physicists in different lunchrooms, libraries, and offices, personal connections were made and the two Bell scientists realized what they might have discovered. The 3 K signal might be real after all, and not due to pigeons or unexplained electrical problems in their antenna, but a signal from the beginning of time. In 1965 Penzias and Wilson finally published their report in the *Astrophysical Journal* and used the title, "A Measurement of Excess Antenna Temperature at 4080 Megacycles per Second," displaying, to the end, extreme caution about the significance of their discovery. A theoretical explanation for the signal was given in the same journal. In 1978 they received the Nobel prize in physics for their important discovery.

The 3 K microwave background signal, which appears to come from all directions in space, has now been firmly established as being due to energy which began its long journey to earth about 300,000 years after the Big Bang occurred. That event created our universe and set space expanding, perhaps forever. When the radio signals began their journey they were light waves, but they have been redshifted into the radio band en route to radio telescopes on earth. At the time the signals originated the universe had cooled down to about 3000 K. The redshift of the source of the 3 K background is so large (the emitting region so far away) that the original light waves have been stretched and diluted to the point of being faint radio signals.

## Beyond the Big Bang—Multiple Universes

Our book has concentrated on the invisible universe revealed, focusing on the discoveries of modern radio astronomy, and it is beyond our scope to enter into a comprehensive description of the theoretical aspects of this science. However, at least one recent development in the understanding of cosmology has to be mentioned in order to explain the observations and relate them to what is known from laboratory measurements on fundamental particles. Our observable universe may be only one of a vast number of universes, all created in the Big Bang. The number of these possible universes is so vast that as far as our imagination is concerned it might as well be infinite.

A new, and more dramatic, variation on the Big Bang theory has recently been suggested. The concept is known as the *Inflationary Universe*. The new cosmology can be related to events that occur in the laboratory, in particular, to the spontaneous creation of particles in a vacuum. According to quantum theory, particles with very little energy can appear from nothing, exist for a brief moment, and then vanish. These *virtual particles,* also known as quantum fluctuations, are observed in laboratory experiments and may also appear spontaneously at any time. The greater the energy, the less time the particle lasts, and a system with exactly zero energy could, in principle, appear from nothing and last forever! Hence we may have a way of creating a universe.

In 1978 three Belgian physicists, R. Brout, F. Englert, and E. Gunzig, suggested a universe-creating process in mathematical detail. Their model predicted that a Big Bang universe could be formed provided a spontaneous pointlike disturbance were created in a previously empty environment. From this seed point a runaway production of mass-energy could result, and this energy would always be balanced by an equal production of negative gravitational energy. The total energy is zero, yet each causes the other to increase. As the universe grows in size, the process slows and stops, but it would generate a flood of hot matter which we later observe as the expanding universe.

The essential point is that under certain conditions matter can be spontaneously created, and given the right initial conditions, this process can go on for a long period of time. The conditions for creating matter exist near a black hole, at what is known as its event horizon. This is the distance from the black hole from within which no light can escape. The physical conditions at the event horizon are so extreme that virtual particles can spring into existence. When they are created they usually disappear again, as they cancel each other out. The physicist Stephen Hawking, however, showed that a black hole can evaporate, with virtual particles appearing near its boundary, some disappearing inside and others leaking into space. Such particles can be created not just at a black hole edge, but at any event horizon. In the Big Bang these event horizons will have been present everywhere, and in vast numbers. They form "bubbles" of ordinary space-time. As soon as a bubble forms, it expands at the speed of light while filling with dense matter by the Hawking process. As the bubble grows, its initial rapid growth gives way to a slower expansion and the creation of matter ceases. However, in the process a universe has been created, a universe which was originally a single bubble in a vast number of such bubbles and now appears to us as an expanding universe.

These modern theories also predict that our universe is not unique. If its creation happened once, a similar creation was possible many times. According to some cosmologists there may be countless universes, each disconnected from all the others. This implies that our observable universe is not all there is, but merely a sample of something far greater, something utterly unknown and unobservable.

Each universe is disconnected from all others because light cannot cross the universes fast enough to communicate about their existence before they have expanded out of sight. According to this new view of the Big Bang, the universe within the observing range of our telescopes had a diameter of 1 cm at the end of the inflationary phase, before the expanding universe we now observe came into existence.

## How Smooth Is Space?

Radio astronomical observations have a bearing on our picture of what happened in that early universe. After the inflationary period each universe would have been very dense and very smooth. But the present-day universe around us is not smooth at all. Great patches of matter, called galaxies, are contained in larger irregular' volumes of space containing clusters of galaxies. Enormous voids exist between them. How did this patchiness come about? Without some gravitational irregularity in the early bubble (which became our universe) there would have been no seeds from which to grow galaxies. Such seeds would be irregularities in the early universe, which should be detectable as slight nonuniformities in the microwave background. Radio astronomers have performed extremely sensitive experiments to discover whether the background shows any tiny variation in its temperature from point to point. So far nothing has been found. The background appears smooth to one part in 200,000, far too smooth to explain how galaxies came to be formed.

## Missing Mass

Another challenging problem in astronomy, and perhaps the most significant cosmological question now being researched, concerns the question of the *missing mass*. Radio observations of galaxies in clusters have confirmed that this is a problem, but have done nothing to help solve it. At issue is the observation that much of space does not contain enough material to hold things together!

We have already discussed gravitational attraction as acting to pull matter inward and internal heat energy as tending to disrupt clouds. In clusters of galaxies each galaxy can be considered as being a particle of a certain mass, a mass which is measured by studying the light from each galaxy. The velocity of the galaxy is known from observations of the spectral lines it emits; for example, the 21-cm hydrogen line. Therefore, the total energy of motion contained in a cluster of galaxies can be calculated, and this energy always turns out to be greater than the gravitational energy available to balance it, so the cluster of galaxies should be flying apart in times short compared with the age of the universe. Yet clusters exist. What holds them together? The discrepancy is known as the missing mass problem.

If more mass, invisible to astronomers, existed in galaxy clusters, they would be gravitationally held together and would last long enough for us to see them as clusters. Astronomers cannot explain why clusters of galaxies form, nor why they survive, once formed. The same problem exists in regard to galaxies and clouds of matter in galaxies. On all scales some matter always appears to be missing, and the larger the volume of space we consider, the larger the missing quantities.

When it comes to the universe as a whole, astronomers would also like to know just how much matter it contains, because this quantity will tell them whether the universe will go on expanding forever or whether it contains enough matter to slow and later reverse the expansion due to gravitational attraction. In the latter case the universe will collapse someday, leading to a catastrophic end to all that exists.

The amount of invisible missing mass may be as much as 97% of all the mass in the universe. A sobering perspective on this mystery is that perhaps this material is not observable by means of electromagnetic waves or particle detectors. Yet there appears to be no other way at present for us to search for the mass. Also, the missing fraction seems to depend on the estimate of the volume of space involved, which depends on knowing the distance to the objects in question. If the distance estimates are systematically incorrect, an option regarded as unacceptable by most astronomers, the missing mass problem might go away. But then we would be saddled with dozens of other problems: for example, to what is the redshift due?

Radio observations of hydrogen in galaxies and motions of gas within clouds have merely confirmed the missing mass problem. No one knows how to solve it. Radio searches for intergalactic hydrogen, which might contribute to a solution, have been largely negative. At this time, there are insufficient data to allow the fate of the universe to be predicted.

## Gravitational Lenses

Another cosmologically interesting phenomenon concerns the gravitational bending of light. This was predicted by Einstein as part of his theory of relativity and provided its first test when, in 1919, the British astronomer Sir Arthur

Eddington measured the deflection of starlight passing close to the edge of the sun during an eclipse. The effect has also been checked by observing the change in position of radio sources situated close to the sun's limb. The apparent shift in position is due to the bending of space by the sun's gravity.

A distant object such as a galaxy can bend the path of light or radio waves from an even more distant object. If the background objects, such as galaxies or quasars, happen to line up with a foreground galaxy and the earth, the effect might be observed.

A gravitational lens concentrates the light (or radio waves) along a line rather than to a single point as glass lenses do for light. A star, galaxy, or quasar also forms a number of images of a background object, with the images displaced from the object's true position if it lies just to one side of the lensing galaxy. The images might also appear brighter than the original object. Several years ago it was suggested that quasars appeared luminous because they were being lensed by foreground galaxies. The general importance of lenses has, however, not yet been proven, but five specific examples of gravitational lens have been discovered.

In 1979, the quasar known as 0957+561 was found to reveal two images 6 arcseconds apart. These appeared to be two quasars whose redshifts and light spectra were identical, something so unlikely as to suggest they were the same object seen in two directions. A faint nearby galaxy was then discovered between them. In 1980 the southern quasar image brightened but the northern one stayed constant until 1982, when it started to brighten while the southern one stayed constant. In 1983 the northern image started to fade. This phenomenon had been predicted. If the double image was due to the lens effect produced by the intervening galaxy, and if the quasar being imaged changed in intensity, then this change was expected to occur at different times in the two images, depending on the different lengths of the light paths en route to earth.

The gravitational lens effect on this source has also been seen at radio wavelengths, although the radio jet for the quasar is seen in only one image. Apparently the position of the radio jet lies sufficiently far away from the lensing galaxy's position that the other focused rays miss the earth.

The nearest lens system observed is the object known as 2237+0305. This is a galaxy approximately $4 \times 10^8$ light-years away, while the quasar whose image is being lensed is 40 times farther. The quasar image is seen in the heart of this galaxy.

Through recent history we have moved from recognizing one earth to many planets, from knowing one sun to recognizing the existence of many stars. We have broadened our horizon beyond our galaxy to include billions of galaxies. Now we shift from a knowledge of one universe created in the Big Bang to a concept that countless universes might exist. We still tend to believe in one form of life—terrestrial. It may only be a matter of time before that prejudice is shattered.

# 15

# On The Search for Extraterrestrial Intelligence

## An Introduction

Human beings have long been fascinated by the possibility of the existence of extraterrestrial intelligence (ETI). As long ago as 50 B.C., Lucretius wrote:

Nothing in the universe is unique and alone and therefore in other regions there must be other earths inhabited by different tribes of men and breeds of beasts.[1]

Today questions about the existence of intelligent creatures on distant planets have taken on a new degree of respectability, thanks to the explosive growth in astronomical knowledge about stars and planets and our recently developed technological capability to communicate with civilizations anywhere in the galaxy. In recent years no book on astronomy appears to have failed to discuss the search for extraterrestrial intelligence (SETI), but in view of our complete lack of knowledge about the nature of ETs, such discussions, despite attempts to present them in a scientific light, appear to be primarily based on speculation, personal belief, and perhaps even hope. The SETI program can only be approached from the point of view of pure exploration, not as a scientifically justified experiment in the usual sense of the word.

It is to astronomers that people turn when the matter of ETI is raised, and radio astrononomers have taken the initiative in keeping the discussion going because communication between inhabited planets is likely to be with radio telescopes. Radio waves are easy to generate and can be beamed over long distances. There is a region in the radio spectrum, around 21-cm wavelength, where the radio emissions from the universe (microwave background, galactic radiation, and radio sources) produce the lowest background intensity and where the absorption produced by intervening matter, including the earth's atmosphere, is at a minimum. Thus the 21-cm band appears to be the natural choice for interstellar broadcasts. Furthermore, the existence of the 21-cm line of interstellar hydrogen gas would be known to other astronomically oriented societies, hydro-

---

[1] Lucretius, *De Rerum Natura*, quoted by M. D. Papagiannis, in *The Search for Extraterrestrial Life: Recent Developments*, Reidel, Dordrecht, 1985.

gen being the most fundamental building block in the universe; therefore, radio astronomers have focused on this band as the one in which the faintest of interstellar calls may yet be heard.

The author was personally involved in SETI in 1971–1973, and at the time was able to demonstrate that no ETs that might be present on planets orbiting certain nearby stars (such as Barnard's star and Tau Ceti) were beaming radio signals at the earth, at least not at 21-cm wavelength. (However, if they were suffering from equipment malfunction or were on a lunch break while I was ready to receive their messages, then the big moment was lost.) This experiment was the first one sensitive enough to have detected civilizations armed with radio telescopes and transmitters of a size we regard as "everyday." Since then, others have spent tens of thousands of hours of telescope time, worldwide, in a continuing and fruitless search for ETI.

At the root of SETI lies an important challenge. Our book has revealed that an enormous variety of radio waves pervade space, but for some reason no signal from ETs has yet been detected. Is this because they aren't out there or because we haven't been lucky yet? There have been a few false alarms, such as the detection of the first pulsar signals (Chapter 11), which were first suspected of being from Little Green Men, and the discovery of the emission from interstellar OH masers (Chapter 10), which had all the hallmarks of a signal from ETI (narrow bandwidth and time variability). These phenomena were relatively quickly accounted for by clever theoretical interpretations, something which gives us cause to wonder whether real signals from ETI might, someday, be similarly accounted for in terms of cosmic exotica.

The current interest in ETI is, however, made more relevant because of the discovery of many types of molecules in space (see Chapter 10). Forty or so different species of organic molecules (Table 10.1), as well as the water molecule, are so pervasive in the Milky Way that life, if it should emerge elsewhere, is likely to be based on the same chemistry that makes us tick. Alien creatures may, chemically speaking, not be all that alien. The real mystery is whether any ETs have evolved a degree of technological sophistication that allows them to search for their neighbors in space, or whether they wish to make their presence known by transmitting radio signals. In this regard, our own technological civilization has only recently evolved to the point of being able to communicate over interstellar distances; hence it is timely to ask whether there are others like us out there.

## Talking across the Galaxy

About a century ago the existence of radio waves was discovered, and before very long they were harnessed for communication over a distance. Today, information-carrying TV, radar, and FM transmissions travel out into space as normal "leakage" from our planet, signals that will indicate to nearby extraterrestrials that we are here. These radio waves may have brushed hundreds of planets in the years since commercial broadcasting and powerful radar systems came into operation. Our planet's radio transmissions have already encompassed a sphere of 40-light-years radius within which ETs could tune into our news broadcasts, football games, and soap operas. But is anyone out there listening, or watching?

In 1959 Guiseppe Cocconi and Philip Morrison suggested that if a highly evolved civilization existed on another planet, it would regard the 21-cm hydrogen line as the choice one for making deliberate contact with other ETs (see Chapter 9). It would not be difficult to transmit a 21-cm signal encoded so that it could not be confused with the natural emissions produced by interstellar hydrogen. Indeed, extraterrestrial civilizations similar to ours may have declared the 21-cm band to be a protected one for the same reasons we earthlings have done so. This band is set aside for the study of galactic and extragalactic hydrogen, and local transmission at this wavelength is forbidden by international agreement. Perhaps ETI will be as wise.

In recent years, sophisticated 21-cm-receiving and data analyzing systems have been built and operated in the search for extraterrestrial intelligence, just in case ETI is signaling. But why would ETs bother? The assumption, and perhaps hope, is that for some reason they are interested in making contact and are taking the initiative to broadcast to anyone ready to receive their transmissions. It is this hope which is the driving force behind the SETI experiments, a hope which, I believe, tells us more about human psychology than about the likely behavioral characteristics of putative extraterrestrials.

As regards the potential for communicating with ETI, the human race did reach a significant milestone in 1976. That was when earth's inhabitants first became capable of communicating with a twin civilization anywhere in the galaxy. If a distant civilization used the equivalent of our largest radio telescope and most powerful radar transmitter, the Arecibo 1000-foot dish (Figure 7.1), we could detect their signals, provided that ETI was pointing its telescope directly at us. The human race now has the capability of talking to an earthlike civilization even if ETs should reside on the other side of the Milky Way, 100,000 light-years away. In that case it would take 200,000 years for a simple, "Hello. How are you?" to get an answer. Hardly a fireside chat, but food for thought.

Natural curiosity drives our search for ETI. The SETI experiments have to be carried out in the tradition of pure exploration, but unfortunately purely exploratory experiments are seldom funded these days. In our scientific era we need to be able to justify experiments in terms of the results that are expected. In the case of SETI we cannot be sure ETs are out there, where to search, how to search, or even whether it is worth searching. We do know that if we do not try we will never know if radio broadcasts from ETI are already bathing our planet.

## The Search Begins

In 1960 Frank Drake, using an 85-foot radio telescope at the U.S. National Radio Astronomy Observatory in Green Bank, West Virginia, searched for radio signals from an alien civilization. His Project Ozma caused a stir in the media, but for the next decade SETI experiments with the world's largest radio telescopes were taboo because many radio astronomers believed (and many still do) that it is a waste of valuable telescope time to perform such searches. Even today it is rare for a SETI program to be granted time on the large telescopes, so ingenious ways to circumvent this limitation have been devised. For example, Paul Horowitz of Harvard designed and built a "suitcase SETI," a portable multichannel receiver which could be carried to any radio telescope in order to search for alien broadcasts. His 64,000-channel device was used with the Arecibo telescope and allowed the data to be immediately analyzed to the point of knowing whether the expected signals were present. None were found. Funded by the Planetary Society, Horowitz plans to use this system, upgraded to eight million channels, to monitor the entire sky once every six months or so, using Harvard's 84-foot radio telescope at Oak Ridge, Massachusetts.

Most of the searches now being done use radio telescopes of this size, and at least one (at Ohio State University) has been dedicated to a continuous search. If Horowitz's new eight-million-channel receiver were attached to the Arecibo telescope, it would be possible to attain in one minute of observation a sensitivity that would have required nearly 100,000 years of full-time operation in Project Ozma's day. This is a measure of the enormous growth in radio-related technology that has occurred over the last quarter-century, a growth that has helped trigger heightened interest in SETI. (Project Ozma used one channel, an 85-foot radio telescope, and a receiver whose internal noise temperature was 350 K. Today the 8,000,000-channel receiver can be used with a 1000-foot radio telescope and a receiver noise temperature of 30 K.)

A NASA-run SETI group at Ames Research Center in California is designing a system capable of simultaneously analyzing 40–100 million channels of information in the hope of someday finding whether ETI is beaming messages into space for all to hear.

## New Initiatives

Sentiment against the radio SETI program, strong within the radio astronomical community during the 1960s and early 1970s, has since lessened. SETI also generated controversy in other communities. For example, on July 31, 1981, the U.S. Senate passed an amendment to the fiscal year 1982 budget which included the following paragraph:

Provided: That none of these funds shall be used to support the definition and development of techniques to analyze extraterrestrial radio signals for patterns that may be generated by intelligent sources.

At the same time, plans were already being discussed within NASA to prepare for the days, not long in coming, when such funds would be reinstated. This moment was perhaps hastened by the publication of a major report concerning the future of U.S. astronomy in the 1980s. Prepared after widespread consultation amongst prominent astronomers, the Field Committee Report (named after its Chairman, George Field) on Astronomy and Astrophysics supported the SETI program as a valid scientific enterprise, although it noted that "SETI is primarily an endeavour characteristic of the exploratory spirit of the Western world." The Committee recommended SETI as a long-term effort, rather than a short-term project in which the microwave radio search was a promising beginning. It also suggested that new approaches be considered, although discussion of alternative approaches outside the radio realm is not obviously encouraged within the scientific community.

The scientific discussion of SETI-related issues was given an aura of greater respectability in 1984, when the International Astronomical Union ran a sympo-sium on the topic of life in the universe and created Commission 51 for members interested in the subject of ETI. The first President of this Commission, Michael Papagiannis, pointed out that even as radio astronomers indulge in the search there are many people who believe that we are already being visited by ETs. "It is not easy to dismiss these contentions either," Papagiannis noted, "since very few scientists are willing to become seriously involved in such investiga-tions."

On April 1, 1986, the NASA SETI program office presented a formal program plan to begin a large-scale search for extraterrestrial intelligence. The plan includes an automated search of the entire sky between 1000 and 10,000 MHz with a resolution of 30 Hz. This survey would be about 300 times more sensitive and have about 20,000 times more frequency coverage than any previous program, and during the program the sky would be scanned at least 30 times. At a cost of about $90 million spread over ten years, this program would be the first systematic study which would provide a reasonable chance of finding radio signals from nearby civilizations, if they exist. The planned project includes a targeted search program which would study nearly 800 solar-type stars within 80 light-years of earth, covering the radio spectrum between 1000 and 3000 MHz with a resolution as small as 1 Hz. The plan emphasizes the study of solar-type stars within 20 light-years of earth, because civilizations orbiting such stars would already have received our radio and TV signals and might, therefore, be making serious efforts to make their presence known to us.

## The Harsh Realities of the SETI Equation

In contrast with the enthusiasm shown by many people to undertake large-scale searches for ETI, astronomers who have reviewed all the data conclude that there is as yet no unambiguous evidence for the existence of planets in the galaxy outside our solar system. That doesn't mean that planets aren't in orbit about other stars; it only means we haven't seen them yet. Hopefully the Hubble Space Telescope will change that picture. The discovery of planets orbiting distant stars is probably the single most important piece of information needed to establish a more rigorous basis for the search for ETI. We also, unfortunately, know nothing concrete about the likelihood that a planet will develop life, or that such life forms will become technological or communicative over interstellar distances.

In gathering together all the unknowns, an elegant way to express the likelihood that there are other civilizations like ours in the galaxy has been created. This is known as the SETI equation, or Drake's equation. We cannot consider all the recent intellectual variations on this theme, which involve more subtle guessing about the motivations and life-styles of ETs, but we can summarize with two tables. Table 15.1 gives the number of galactic civilizations capable of communicating over interstellar distances and Table 15.2 gives the average distance (in light-years) between them.

The basic SETI equation is given by:

$$N = R^* f_p n_e f_l f_i f_c L,$$

where $R^*$ is the rate of star formation in the galaxy, in number of stars per year; $f_p$ the fraction that have planets; $n_e$ is the number of ecologically suitable planets orbiting a typical star; $L$ is the lifetime of a typical communicative ET society. The other three $f$ terms are fractions indicating the percentage of the ecologically suitable planets which give rise to life ($f_l$), the percentage of those that produce intelligence ($f_i$), and the fraction of those that, in turn, become communicative ($f_c$) over interstellar distances.

The stellar birthrate is approximately given by the total number of stars in the galaxy ($2 \times 10^{11}$) divided by the age of the galaxy ($2 \times 10^{10}$ years) or about 10/year. If we assume that each star has one planet ecologically suitable for life (e.g., the sun has only the earth in this category), probably a very optimistic estimate, we can find the number, $N$, of civilizations to talk with, depending on the values estimated for each of the fractions and the lifetime. In the tables the vertical columns consider several options in which the three fractions have been combined (i.e., $f_3 = f_l f_i f_c$). Thus $f_3 = 10^{-3}$ means that a civilization capable of communication over interstellar distances emerges on

**TABLE 15.1.** Where is everyone?
*The number of civilizations capable of interstellar communication.*

| Lifetime (years) | Fraction of planets producing communicative ETs | | | |
|---|---|---|---|---|
| | $f_3 = 10^{-2}$ | $f_3 = 10^{-3}$ | $f_3 = 10^{-4}$ | $f_3 = 10^{-5}$ |
| | *The number of civilizations* | | | |
| $10^3$ | 100 | 10 | (1) | (0.1) |
| $10^4$ | 1000 | 100 | 10 | (1) |
| $10^5$ | 10,000 | 1000 | 100 | 10 |
| $10^6$ | 100,000 | 10,000 | 1000 | 100 |
| $10^7$ | 1,000,000 | 100,000 | 10,000 | 1000 |

(  ) indicates meaningless number—us or no one!

**TABLE 15.2.** Where is everyone?
*The distance in light-years to the nearest civilization capable of interstellar communication.*

| Lifetime (years) | Fraction of planets producing communicative ETs | | | |
|---|---|---|---|---|
|  | $f_3 = 10^{-2}$ | $f_3 = 10^{-3}$ | $f_3 = 10^{-4}$ | $f_3 = 10^{-5}$ |
|  | *Distance in light-years to the nearest civilization* | | | |
| $10^3$ | 8000 | 25,000 | (80,000) | (250,000) |
| $10^4$ | 2500 | 8000 | 25,000 | (80,000) |
| $10^5$ | 800 | 2500 | 8000 | 25,000 |
| $10^6$ | 250 | 800 | 2500 | 8000 |
| $10^7$ | 90 | 250 | 800 | 2,500 |

(  ) indicates meaninglessly large number—outside the galaxy.

one in a thousand ecologically suitable planets. Choose a value for this fraction and then choose your favorite value for lifetime, $L$. The number you find in Table 15.1 is then the number of civilizations in the galaxy, while the corresponding entry in Table 15.2 is the distance in light-years to the nearest neighbor.

The pattern in the tables should allow you to easily extend the entries to the right, where the likelihood of companions in the galaxy quickly drops to zero. The numbers in Table 15.2 refer to the time for a one-way signal to get from them to us. Double it to figure out how long it takes to get an interstellar chat going. Obviously, unless typical lifetimes are large and a high fraction of likely habitats spawn civilizations, we will have a terrible time conversing with anyone. In fact, if lifetimes are 100,000 years or less, there really is no one to talk to unless we are willing to spend millenia in efforts to engage in startalk.

The typical lifetime of a communicative civilization is the greatest unknown which bedevils a scientific approach to SETI. This time period has to be large if there is to be any chance of contact. It is possible that ETI has long since emerged and evolved on, and disappeared from, other planets. Estimates for the average lifetime of a civilzation are clearly related to questions of survival, extinction, and evolution.

Evolution probably occurs throughout the universe and will manifest itself on every planet on which life claims a foothold. A million years from now our species will surely be very different from anything *homo sapiens* can imagine today. If we allow for this possibility, how do we search for ETI? Can we believe that a million years from now we will be sufficiently similar to our present form to be a technological civilization using radio telescopes to communicate with ETs? This is not a popular question with the SETI proponents, who must answer in the affirmative. However, if evolution continues, it seems more likely that a million years from now our species will be as unrecognizable to us in terms of its form and function as we are to the apes. We cannot predict what will happen, but surely evolution will continue after the twentieth century, and who knows what that might bring?

## The Search Goes On

And so the SETI program goes on, because if we do not search for ETs we will never know if are out there. But do not expect quick results!

Many other fascinating questions are being discussed in regard to this topic. For example, it has been argued that because there are no ETs on earth today, they do not exist elsewhere in the galaxy. It is a clever argument, based on simplistic assumptions. The proponents neatly demonstrate that any technological society could, if it wished, easily colonize the galaxy in a matter of a few million years; thus they should already be here. Why aren't they? Answer— because they aren't out there. Ignoring the fact that we cannot prove that there are no ETs on earth today, we might instead conclude that evolved extraterrestrials abandoned the colonizing mentality.

What about the consequences of contact? Genetic engineering draws close scrutiny in view of the profound changes it may wreak upon future generations, but surely if contact with ETI is established it may also have a profound bearing on our collective future. Some radio astronomers have argued that contact with ETI may be the most profound event that has ever occurred in the history of the human species. Should we be concerned? Should we discuss it now, just as genetic engineering is being closely monitored? I believe it deserves consideration.

In summary, there are many radio astronomers who are fascinated by the fact that there is a subtle silence amongst the stars. It is significant that we can see into the hearts of quasars, listen to pulsars, and sample the chemical fragrances inside molecular clouds, but have not received a call from ETs. Are they not out there? Or is it that our assumptions and beliefs about the origin and evolution of life and about the nature of technological civilizations, and consequently about how to seek out ET, are wrong? The SETI proponents assume, as they must, that ETI is enamored with technology and that ETs love radio telescopes as much as we do. Finally, they must assume that ETI wishes to make itself known to newcomers on the galactic scene. But perhaps ETI knows something we don't  . . .

# Part VI

## Radio Astronomy Review: Past, Present, and Future

# 16

## Musings on the Evolution of a Science

### Early History

A century ago, in 1888, Heinrich Hertz constructed the first radio transmitter and receiver. A spark created in his "transmitter" caused a response in the "receiver" placed some distance away. An invisible form of radiation had carried energy through intervening space. Hertz was able to demonstrate that these radio waves, as they came to be called, were a phenomenon very similar to light, and today we know that both are forms of electromagnetic radiation. Nearly 50 years later, in 1933, radio waves from the Milky Way were discovered, and now, a century after Hertz, we constantly receive radio waves created by "sparks" in distant galaxies and quasars.

Four distinct phases can be recognized in the story of the evolution of radio astronomy. Phase I, up to 1950, is characterized by an initial identity crisis as the new science tried to define itself. It was a time during which radio engineers made a series of startling discoveries, very few of which could be understood in terms of previous astronomical knowledge. Professional astronomers took relatively little notice of radio astronomy during this period. The second phase, 1950–1960, was marked by confusion while controversy raged as to the meaning of the new discoveries. Optical astronomers barely began to recognize that radio astronomy existed. During phase III, 1960–1980, an enormous number of new discoveries were made and significant, very welcome progress in theoretical understanding was achieved. The present Phase IV, which began around 1980, is the age of consolidation and clarification and is characterized by an extraordinary improvement in the radio astronomer's ability to see radio sources more clearly. Relatively few fundamentally new discoveries, characteristic of Phase III, appear to occur.

### Phase I—Birth

Karl Guthe Jansky, the father of radio astronomy, was employed by Bell Laboratories, which, in 1927, introduced the first transatlantic radiotelephone. For a mere $75 you could speak for three minutes between New York and London, but the radio links were terribly susceptible to electrical interference. The first system operated at the extraordinarily low frequency of 60 kHz (that is, at the very long wavelength of 5 km) and in 1929 a switch was made to short waves whose frequencies were in the range 10–20 MHz. But the new telephone links were also very susceptible to electrical disturbances of an unknown nature, which plagued the connections. Jansky was assigned the task of locating the source of the interference. To carry out his studies he built a rotatable antenna (Figure 16.1) operating at 20.5 MHz and by 1930 began regular observations. In 1932 he presented his first report. Local and distant thunderstorms were two sources of some of the radio noise and a third source was, ''a very steady hiss type static, the origin of which is not yet known.'' When Jansky became convinced he had picked up radio waves from space, he must have enjoyed what few people ever experience—the thrill of finding something never before seen by anyone, anywhere. That is part of the challenge, joy, and excitement of doing scientific research.

During the next year he was able to demonstrate that the source of the signals was outside the earth and presented a report entitled ''Electrical Disturbances Apparently of Extraterrestrial Origin.'' And so radio astronomy was born.

Fifty years later, at the National Radio Astronomy Observatory in Green Bank, West Virginia, distinguished radio astronomers gathered to celebrate the fiftieth anniversary of Jansky's discovery. A report, entitled *Serendipitous Discoveries in Radio Astronomy,* which presents the human side of the birth and growth of this science, grew out of that meeting.[1]

''Serendipity'' is a term coined by Horace Walpole, the writer and historian, who used it to refer to the experience of making fortunate and unexpected

**FIGURE 16.1.** A replica of Karl Jansky's rotating turntable antenna erected at the National Radio Astronomy Observatory, in Green Bank, West Virginia. (National Radio Astronomy Observatory.)

discoveries, following the fairy tale about the three princes of Serendip (an old name for Ceylon). Serendipitous discoveries are those made by accident, but also by wisdom, for no one can make an accidental discovery unless that person is capable of recognizing that something of significance is occurring. Jansky was such a person.

In January 1934, in a letter to his father, Jansky wrote:

Have I told you that I now have what I think is definite proof that the waves come from the Milky Way? However, I'm not working on the interstellar waves anymore.[1]

His boss had set him to work on matters of more immediate concern, matters which were

. . . not near as interesting as interstellar waves, nor will it bring near as much publicity. I'm going to do a little theoretical research of my own at home on the interstellar waves, however.[1]

Unlike Grote Reber after him, Jansky did not take his interest in his new discoveries to the point of building his own antenna so as to pursue his explorations over the weekends. Jansky's boss, who ruled with an iron hand, was later to encourage him to write another report, and in 1935 Jansky interpreted the sky waves as coming from the entire Milky Way. But he did not know why and suggested that either a lot of stars were contributing or something in interstellar space was the cause. He realized that if the waves were due to stars he should have picked up the sun. The Milky Way happens to reach maximum brightness, as seen from the surface of the earth, close to Jansky's chosen frequency. It is brighter at still lower frequencies, but those radio waves do not penetrate the ionosphere, the electrically conducting layer hundreds of kilometers up in the atmosphere. Furthermore, the ionosphere experiences daily changes of its characteristics and in the daytime blocks out the sun; thus it blinded Jansky's antenna to its radiations. The mid-1930s were also a time of sunspot minimum, which meant that the ionosphere was transparent to 20 MHz at night. Had Jansky been observing at sunspot maximum, the ionosphere would have blocked out all 20-MHz radio waves from space and he would not have discovered the signals from the Milky Way.

Jansky did not pursue his discoveries any further because there were other projects to be done and "star noise could come later," he was told by his employers. It was to be years before significant follow-up work began. A few astronomers in the U.S. and Europe had become aware of Jansky's work, but any plans to pursue his discoveries had to be shelved when World War II broke out. In any event, most astronomers knew absolutely nothing about radio receivers and antennas, so how could they get involved? These astronomers were not equipped with the skills necessary to get involved in radio astronomy. After the War it was mostly radio physicists who launched the new science, and they had to learn astronomy in the process.

Jansky was to write to the famous physicist, Sir Edward Appleton:

If there is any credit due to me, it is probably for a *stubborn curiosity that demanded an explanation* for the unknown interference and led me to the long series of recordings necessary for the determination of the actual direction of arrival.[1] (my italics)

Such *stubborn curiosity* is the hallmark of good scientists. Jansky trusted his data and continued his measurements for confirmation. His persistence led to the discovery of the long-term variations which showed that the source of the static was in astronomical space.

The story of radio astronomy is replete with apparently amazing cases of fortuitous discoveries, but such discoveries required more than good luck. They required prepared minds and dedicated effort to follow up on what might at first have seemed to be a preposterous new observation.

## Caught between Two Disciplines

In 1933 John Kraus, then at the University of Michigan, attempted to detect the sun at 15-mm wavelength, using a searchlight reflector to focus the radio waves. He failed because the receiver was not sensitive enough, but his was the first use of a reflector-type radio telescope. At the Serendipity meeting Kraus stated that serendipitous discovery only occurs as the result of "being in the right place with the right equipment doing the right experiment at the right time." Another noted astronomer, R. Hanbury Brown, added that the person should "not know too much," otherwise the discovery may not be made!

This summarizes a very interesting phenomenon. Many research scientists, especially the theoretically inclined, "know" so much that their chances of making a lucky or creative discovery may be severely curtailed. If we know too much, our vision is sometimes narrowed to the point where new opportunities are not seen. Jansky knew a little astronomy, but not enough to get in his way and cause him to reject the possibility that cosmic radio waves might be real.

Grote Reber, professional engineer and radio ham in his spare time, was one of the few people who recognized the interesting implications of Jansky's discovery. Reber was certainly not hampered by any astronomical prejudices about whether or not the cosmic radio waves could exist. He was interested in verifying their existence and followed up on Jansky's work. To this end, Reber built the world's first steerable radio dish antenna (Figure 16.2) in his backyard and mapped the Milky Way radiation during the period 1935–1941. He recently pointed out that the new field of radio astronomy was caught between two disciplines. Radio engineers didn't care where the radio waves came from and the astronomers

. . . could not dream up any rational way by which the radio waves could be generated, and since they didn't know of a process, the whole affair was (considered by them) at best a mistake and at worst a hoax.[1]

**FIGURE 16.2.** The pioneer radio astronomer Grote Reber, visiting his radio telescope now on public display at the National Radio Astronomy Observatory, in Green Bank, West Virginia. (National Radio Astronomy Observatory.)

The very essence of research is that once an observation is made it requires some understanding and interpretation in order to formulate a plan for making further observations. It was initially very difficult for astronomers, entirely ignorant of radio technology, to interpret or understand the significance of Jansky's or Reber's epoch-making discoveries.

Jesse Greenstein, of Caltech, one of the few astronomers who did get involved before World War II, summed up the dilemma confronting the astronomer of those prewar days:

I did not say that the radio astronomy signals would go away someday, but I didn't know what next to do.[1]

How could anyone know what next to do? The mystery of where the radio waves originated was a profound one, not easily solved. Significant new technologies had to be combined with astronomical knowledge in order to do radio astronomy research. If the science was to flourish, either astronomers had to learn about radio engineering or radio engineers had to learn astronomy. The new science therefore grew slowly. The intrusion of World War II may actually have enormously speeded up its growth because of the intense research in radar techniques, which led to the very rapid development of precisely those radio antennas and receivers which radio astronomers were to require for their work and which were readily available to them after the war.

## Postwar Years—Radar Everywhere

England, Australia, France, the Netherlands, the U.S., and Canada were the important centers for postwar radio astronomy. The radio engineers and physicists drawn into radar research during the war became the first generation of professional radio astronomers. The equipment they used to launch their research work was scrounged, begged, or borrowed from military surplus.

In 1946 in Canada, Arthur Covington, of the National Research Laboratories, began regular observations of the sun at 10.7-cm wavelength, a choice dictated by the availability of surplus radar components. For decades this work has provided the standard data for anyone interested in knowing how active the sun is on any given day. The solar radio data showed that the sun's radio brightness is directly correlated with the 11-year sunspot cycle and also revealed that the radio active regions must be at temperatures of over 1,000,000 degrees.

Radio astronomers in the Netherlands did their early work with a German war surplus radar (Würzburg) dish. Since 1944—when H. C. van de Hulst, a graduate student working with J. H. Oort, had given a talk on how radio observations might contribute to our understanding of the universe—the Dutch had focused their attention on the 21-cm hydrogen line, which led to its discovery in 1951.

The Cambridge radio astronomy effort, under Martin Ryle, made heavy use of two Würzburg dishes, which were used to accurately locate some of the strongest radio sources in the sky with sufficient accuracy that optical identifications could be made. In 1948 Ryle and F. G. Smith discovered Cas A. Then Smith succeeded in improving the radio measurements to the point where an accuracy of one minute of arc allowed the optical astronomers on Palomar mountain to photograph the position and discover the filamentary remains of the supernova (Figure 8.5). The position of Cygnus A was also measured accurately enough to lead to its identification with a very faint, distant galaxy.

During the later phases of the war, radar antennas in England had been pointed above the horizon to detect incoming V2 rockets, and in the process accidentally detected echoes from meteor showers. As meteors burn up in the atmosphere, they produce ionized trails which reflect radar signals. This discovery interested Bernard Lovell, of the University of Manchester, who wanted to find similar echoes from cosmic rays striking the atmosphere. As a pioneer in aircraft radar development, Lovell had access to surplus radar equipment which the University allowed him to park at their botany research station at Jodrell Bank, south of Manchester. (A peculiar coincidence: Jansky lived in a town called Red Bank, New Jersey. The U.S. National Radio Astronomy Observatory is located at Green Bank, West Virginia. Lovell set up shop at Jodrell Bank,

England. This surfeit of banks is not a reflection on the profession's remunerative benefits!)

Lovell's observations revealed no cosmic-ray echoes, only more and more meteor trails, and for years meteor astronomy was a focus for research at Jodrell Bank. As the radio antennas grew in size, so did their potential for doing radio astromomy. Lovell subsequently propelled Britain into the forefront of the science by masterminding the construction of what was for many years the world's largest fully steerable radio telescope (the 250-foot Mark I). Completed just before the world's first artificial satellite, Sputnik I, was launched in 1957, the Mark I was the only radio telescope in the world capable of picking up radar echoes from the satellite's carrier rocket and played an important role in stimulating the U.S. to get more active in radio astronomy and to develop a more effective radar for national defense.

## The Southern Skies

Observation of the southern skies fell to the Australian radio astronomers led by J. L. Pawsey, studying the sun, and J. G. Bolton, studying other radio sources, who also began by using surplus radar equipment. They invented a neat trick to make their radio antennas work more effectively. The resolution obtainable by a radio telescope depends on the diameter of the dish. If two antennas are used and separated by some distance, the resolution of the combined antennas (known as an *interferometer*, see Chapter 17) is determined by the distance between them. Instead of building two antennas and spacing them hundreds of yards or miles apart, Bolton's team placed an antenna on top of a cliff by the ocean. To long-wavelength radio waves the ocean's surface acts like a mirror. From the radio source's point of view, the clifftop radio telescope consisted of two antennas, one on top of the cliff and one apparently some distance below it, seen reflected off the water. These two antennas, one real and one a reflection, acted together as an interferometer.

With this ingenious device, capable of 10 arcminutes of resolution, the Australians made some of the most important observations in early radio astronomy. They discovered that enhanced solar radio emission was associated with sunspots (in 1946) and that the temperature of the quiet sun was 1,000,000 degrees, and they observed the first solar burst (1947). They confirmed the position of the Cygnus A radio source (1947) and found several new radio sources, which helped arouse the interest of optical astronomers in the new science. They discovered the Taurus A radio source in 1947, and their accurate position measurements facilitated its identification with the Crab nebula (1948). Then Centaurus A and Virgo A were added to the very short list of radio sources associated with galaxies.

Observations of the radio emission from our galaxy (in 1952 and 1953) led to the discovery that there was a bright radio source in the constellation of Sagittarius. At the time it was believed that the galactic center was about 32°

**FIGURE 16.3.** The 210-foot-diameter Parkes radio telescope at Parkes in Australia. The inner dish is solid, while the outer parts are made of fine mesh. (Commonwealth Scientific and Industrial Research Organization.)

away from this radio source, but subsequently, by international agreement in 1955, the location of Sgr A was taken to define the true galactic center.

In later years the large radio telescope at Parkes (Figure 16.3), near Sydney, was used in fundamental discoveries related to the polarization of radio sources and polarization of the galactic radiation. The Parkes surveys of radio sources became the basis for early cosmology studies and led to some very intense debate between Australian and Cambridge radio astronomers over the interpretation of the data.

## Who Could Have Guessed?

By the time 1950 began, the radio discoveries had barely made an impression on astronomers. The new breed of radio astronomers had discovered radio waves from the Milky Way and the sun and had managed to locate several radio sources which had been optically identified. However, the picture appeared very confusing. Nearby galaxies, such as M31, were very faint radio sources, but some very distant galaxies, such as Virgo A and especially Cygnus A, were powerful emitters of radio signals. Centaurus A was associated with the galaxy NGC 5128 and was clearly not at all well behaved (see Chapter 3), and no accepted theory yet existed to explain the radio emissions. In fact, around 1950

. . . radio astronomers were greatly impressed by the almost total lack of connection between radio observations and the visual sky. It did not seem impossible then that there were two separate kinds of celestial objects, each requiring distinct research techniques.[2]

With regard to the problem of explaining the existence of the newly discovered radio sources, Greenstein commented:

No rational explanation that explains the weak (radio) emission from the brightest nearby galaxy, the Andromeda Nebula, can also apply to the faint distant radio source Cygnus A. You have to break down the prejudice that the world is pretty much as you know it, and begin to think of a world which is not like the world you understand.[1]

Breaking through prejudice is something which has frustrated many a scientist (as well as philosopher, politician, or layperson). Oddly enough, the holding of scientific beliefs creates a secure structure of apparent prejudice within which to operate, in order to make new discoveries which will then force a change in those very prejudices (or beliefs).

Who, back then, could possibly have guessed at the amazing scenario we now know accounts for the cosmic radio waves. Radio signals from the Milky Way are produced by cosmic-ray electrons spiraling about interstellar magnetic fields. In the thirties and forties no one knew that interstellar space contained cosmic rays or that there were magnetic fields between the stars. The existence of emission nebulae was known, but attempts to explain the cosmic static by filling interstellar space with thermal electrons or radio-emitting stars or emission nebulae did not work because that required gas temperatures of millions of degrees, not merely the 10,000 K known to exist in emission nebulae.

The cosmic-ray connection was also interesting. At the time, cosmic rays were defined as protons (but not electrons) from space which struck the earth. Cosmic-ray physicists didn't concern themselves too much about the origin of the cosmic rays nor about what had happened to the electrons. The researchers were mainly interested in studying the composition and physical properties of the particles that did reach their detectors. The absence of electrons was noted, but who would have thought that the electrons didn't reach the earth because they had wasted their energy radiating radio signals in interstellar space?

After the war, Enrico Fermi proposed that cosmic-ray electrons could be accelerated in interstellar space, provided magnetic fields were present, but it wasn't until 1951 that the fields were discovered through the observation of the polarization of starlight by dust grains aligned by these fields. Later, when supernova remnants were recognized as strong sources of radio waves and their polarization measured, astronomers realized that cosmic rays originated in supernovae. Neutron star engines (pulsars) inject energetic electrons into the supernova remnants. These electrons, spiraling about magnetic fields, cause the supernova remnants to shine and then they go on to make the entire Milky Way glow with radio energy. Who, back in 1950, could ever have dreamt up something so outrageous?

## Phase II—Identity Crisis

During the 1950s more radio sources were discovered and cataloged, and arguments raged as to what the new data meant. The first generation of large radio dishes, the Jodrell Mark I (1957) and the NRAO 300-foot (1962), were years from completion. Receiver technology was still relatively crude and radio observations were hampered by receiver noise in efforts to detect faint radio signals from space. By the late 1950s the science appeared to be in a state of relative confusion. David O. Edge and Michael J. Mulkay, who have traced the early development of radio astronomy in their book *Astronomy Transformed*, observed that by 1958

. . . we are . . . at a time of maximum uncertainty and confusion in the history of work on radio sources. Agreement between the two major groups engaged in survey work (in Australia and Cambridge) is minimal, and the status of many of the observations is radically in doubt.[2]

Argument raged, and regarding the general state of radio astronomical knowledge at the time, these authors ask:

. . . what was achieved, by 1958 . . . ? A handful of optical identifications, of an odd assortment of objects, "normal" and "abnormal," a suggested mechanism for radio emission from some of these (this being very largely the work of optical astronomers and theorists); a growing realization (many having already realized it quite early in the fifties) that the majority of radio sources must be extragalactic . . . ; catalogs of sources numbering, for all the hopes, merely hundreds, and those still the subject of controversy; some (but not many) radio diameters, spectra, and a few polarization measures; cosmological claims radically in doubt; source counts in complete disarray . . .[2]

Radio astronomers, who were not even considered by traditional stargazers to be astronomers at all until the late 1950s and early 1960s, had clearly stumbled into a new universe, an invisible universe. Much like a blind man has to learn his way around his world, the radio astronomer not only had to develop new ways of sensing what was out there, but had to invent methods for communicating what it was he discovered. After careful communication with optical astronomers, the radio astronomer attempted to infer whether his observations related to something known to other astronomers or whether he was sensing a completely different universe.

The authors of *Astronomy Transformed* have also suggested that radio astronomy went through several stages. The first stage began with the discovery of radio waves from objects like the sun and the early exploration of these discoveries. There was a sharing of information between several groups and by the end of this stage (early 1950s) it was recognized that there were several very astronomical lines of inquiry involved:

During the ensuing stage radio astronomers publish increasingly in optical journals, join optical (astronomy) societies . . . and come to hold joint conferences with optical astronomers. Essentially a bond is formed with the "real" astronomers. The radio technology is developed so that good data, which make sense and are repeatable, are generated.[2]

## Phase III—Discovery

It wasn't until the 1960s that the bond with "real" astronomers began to be forged on a large scale, following stunning new discoveries made possible by enormous improvements in receiver technology and the construction of large reflector-type radio dishes and larger interferometers. These contributed to the radio astronomer's ability to measure radio source positions with greater accuracy, sufficient to force the attention of the general community of optical astronomers. The days when an old-timer at a meeting of the Royal Astronomical Society could ask, "What is this newfangled wireless astronomy?" were past.

The 1960s also saw the transformation of radio astronomy into a "big science," which brought with it a remarkable period of exciting new discovery. Research at the forefront was, however,

. . . only open to those groups with sufficient expertise to develop the complex techniques required and with sufficient repute to attract extensive financial support from government and from industry.[2]

It should be stressed that radio astronomy growth during Phase II largely bypassed the U.S., and only when Phase III dawned did U.S. radio astronomers begin to catch up.

The 1960s and early 1970s saw the discovery of quasars, pulsars, radio source polarization, complex interstellar molecules, interstellar masers, radio stars, bipolar flows, radio jets, and extragalactic molecules, and the first measurement of the interstellar magnetic field strength. Those years also saw the solidifying of the theoretical understanding of the emission mechanisms involved in thermal and nonthermal radio sources, while explanations for the maser mechanism as well as pulsar radiation were quick to develop.

According to Edge and Mulkay:

Stage three is characterized by a growing concern with astrophysical problems, arising largely from the major discoveries of quasars and pulsars and from the advent of new approaches like those of ultra-violet, X-ray, and infra-red astronomy. By this stage radio methods have become an established part of astronomy.[2]

By the mid-1960s and certainly at the end of that decade, it was firmly demonstrated that the universe was not as quiet as had long been assumed. The universe is wracked with violence on all scales from exploding stars through exploding galaxies and quasars to violence on the scale of the universe itself, the Big Bang.

From the point of view of growth, and availability of funds to drive this growth, the period 1960–1975 might be called the Golden Age of radio astronomy. That was also when radio astronomy terms such as quasar and pulsar entered the vocabulary of the man in the street.

# Phase IV—Consolidation and Clarity

Our book has mostly been about the discoveries made during Phase III. Enormous detail in galactic and extragalactic radio source structures are now being revealed during Phase IV, the age of the very large radio telescope systems, such as the VLA, and the Cambridge and Westerbork synthesis telescopes (see next chapter). Since 1980 radio astronomers have looked more and more closely at those types of radio source already known to exist. The dramatic radiographs of Cygnus A and Cassiopeia A, while revealing to us what radio sources look like, are not observations of new classes of objects, however. The discovery of new objects in the universe, such as quasars and pulsars clearly were, has apparently ceased. This is a common phenomenon in astronomy. Soon after new techniques or telescopes are brought into use the discovery rate peaks and then drops off as the cream is skimmed off the top, so to speak. Then the newly discovered objects or phenomena become subjects for routine study.

Will radio astronomy ever again enter a period of high excitement where new discoveries, of significantly different types of objects or phenomena, dominate the headlines? It seems unlikely. The very high-resolution radio telescopes now being used or planned (see next chapter) will help us understand many of the secrets of the radio universe, but unless new windows into the universe are opened up (and there are a few left for radio astronomers—the millimeter wavelength region, for example), we may expect the flood of significant new discoveries to dry to a trickle. Radio astronomy, in years to come, will mature and become routine. But far more detailed, and hence still exciting, observation may become its hallmark. This phase may make it difficult for this ''big science'' to continue at the level of funding it now enjoys.

The full history of radio astronomy is presented in several excellent books listed in the bibliography. We have touched upon a few anecdotal aspects which hopefully give the reader some feel for what was involved in the growth of this science. Serendipitous, or apparently accidental, discoveries have played their part, but no one can make a discovery without being prepared for it. As Louis Pasteur said, ''In fields of observation, chance favours only the mind that is prepared.'' So it has been in radio astronomy. The important discoveries were made because the right person had the right equipment and was doing what turned out to be the right experiment at the right time, even if the results turned out to be unexpected. Also that person did not know too much, was not so sure of his or her earlier prejudices or beliefs as to be blinkered. Then, only after the scientist dedicated a lot of ''stubborn curiosity'' to the problem, was progress made.

## References

1. K. Kellermann and B. Sheets (eds.), *Serendipitous Discoveries in Radio Astronomy*. National Radio Astronomy Observatory, Green Bank, West Virginia, 1983.
2. D. O. Edge and M. J. Mulkay, *Astronomy Transformed*. Wiley Interscience, London, 1976.

# 17

## Radio Telescopes—The Present

### Bigger and Better

The quest to see the radio universe more clearly has led to the construction of bigger or better radio telescopes and receivers which together satisfy three criteria: greater sensitivity, lower noise, and higher resolution.

The larger the physical (collecting) area of a dish-shaped radio telescope, the greater its sensitivity and the fainter the radio signals it can detect. The 1000-foot Arecibo dish in Puerto Rico (Figure 17.1), completed in 1963, covers 20 acres, larger than the combined area of all the other radio telescopes in the world. It is built in a natural depression in the hills south of Arecibo, and relatively little work was required to scoop out the terrain so that the 1000-foot-diameter reflecting surface could be fitted into this hole in the ground. The reflecting surface cannot be moved, but radio sources can be tracked by moving the antenna at the focus along a specially designed support structure.

The smoothness of the surface of any telescope mirror, whether it be an optical or radio telescope, determines how good a reflector it is. The better the reflector, the more of the gathered energy is accurately brought to a focus and the less goes to waste by bouncing off in random directions. The surface irregularities have to be smaller than one-eighth of the observing wavelength in order for the mirror to perform well. A radio dish may look rough to the eye but to a radio wave it appears as a shiny mirror. After being resurfaced in the mid-1970s, the Arecibo dish operated down to 6-cm wavelength, which means that over its entire surface the irregularities are smaller than 1 cm. This was achieved by accurately adjusting each one of 38,778 panels to an accuracy of a few millimeters.

The telescope is extensively used for planetary radar, pulsar work, and the study of the hydrogen and OH in distant galaxies, for which its enormous collecting area makes it by far the most sensitive radio telescope in the world. It is able to pick up fainter signals than any other system. However, all large radio telescopes are potentially capable of detecting the microwave background, so they can all see at least one thing equally far out into space: the boundaries of the observable universe.

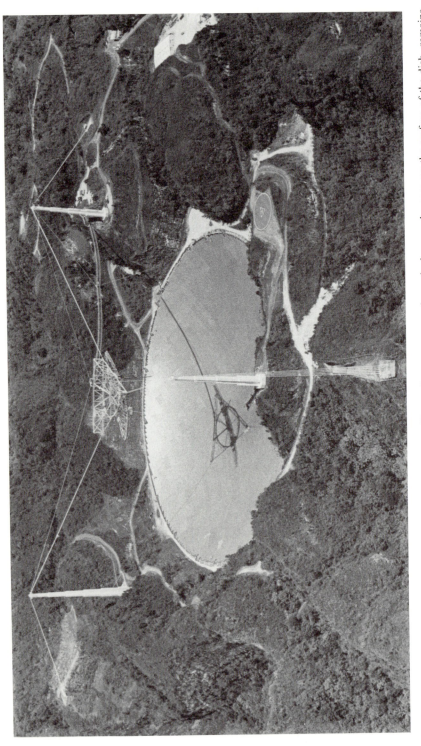

**FIGURE 17.1.** The 1000-foot radio telescope situated in the hills south of Arecibo in Puerto Rico and operated by the National Ionospheric and Astronomy Center of Cornell University. Despite the enormous span of cables suspended from the three support towers, the central antenna structure, whose shadow can be seen on the surface of the dish, remains steady in a strong wind and has been designed to withstand hurricanes. (Arecibo Observatory.)

It seems very unlikely that anyone will soon build a single dish radio telescope larger than the 1000-footer on earth, because of the cost and the engineering difficulties. There are physical limits on how large a single dish can be built. A steel structure, if made too large, buckles under its own weight. Although 500- and 600-foot-diameter, fully steerable radio telescope designs sat on drawing boards for years, they were never constructed due to engineering limits and, of course, severe financial constraints. Size alone, however, is not all there is to detecting faint radio sources.

## Low-Noise Receivers

The radio amplifiers, or receivers, connected to the antenna at the focus of the dish-shaped reflector have to be of extraordinary quality to match the effectiveness of the radio telescope. Internal noise generated in the radio amplifiers, and in the cables or other hardware that connect the antenna to the receiver, compete directly with the faint radio whispers from space. If the internal noise levels are too high, faint radio sources will not be detected. In the early days of radio astronomy, vacuum tube receivers were connected to Yagi antennas (which are similar to conventional TV-type antennas located on rooftops). The internal noise temperatures of vacuum tube amplifiers were the equivalent of thousands of degrees. Then, in 1960, a device called a parametric amplifier became available. Its noise temperature was first 350 K, and then later development reduced it to around 200 K. This meant that the sensitivity of radio telescopes suddenly improved tenfold, because the sensitivity depends directly on the noise temperature of the receiver. Modern radio telescopes use extremely low-noise receivers to produce overall noise temperatures of 20 or 30 K, another tenfold improvement over parametric amplifiers.

Radio receiver technology is unlikely to lower these noise temperatures very much further because of limitations that have little to do with engineering. The 3 K microwave background sets the absolute limit, and atmospheric radiation contributes another 6 K. A lower bound of 10 K is believed to be a realistic goal for radio telescopes in space.

## Interferometers

In order to see more clearly, the radio astronomer needs, above all, high resolution. Single-dish radio telescopes provide only fuzzy images of radio sources. The vintage 300-foot diameter radio telescope (completed in 1962) at the NRAO in Green Bank, for example, has a resolution of 10 minutes of arc at 21-cm wavelength. Any object it studies, no matter how small the object may be on the sky, whether one second or one minute of arc across, is made to look as if it is 10 minutes of arc in size.

For years, radio astronomers have concentrated their efforts on ingenious methods for producing far greater resolutions without building huge dishes.

For example, they combine antennas to simulate larger ones. Two or more radio dishes, separated by some distance, are electronically linked to form what is called an *interferometer* (Figure 17.2). When the radio signals from the two antennas are added they can either reinforce or interfere, that is, cancel each other out. The operation of an interferometer is most easily pictured if you imagine a simple wave entering both antennas, as shown in Figure 17.2. When the two waves arrive simultaneously, they reinforce each other, but if one arrives slightly later, delayed by, say, half a wavelength, the minimum of

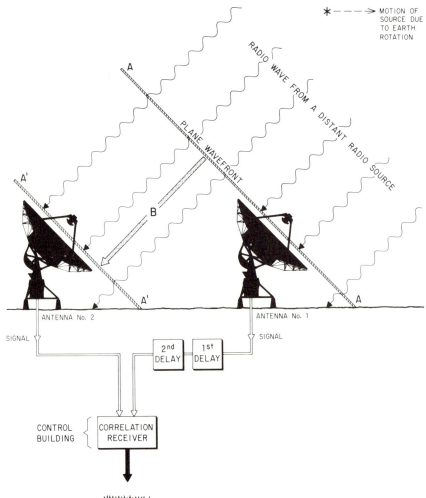

**FIGURE 17.2.** How an interferometer operates (see text). (National Radio Astronomy Observatory.)

one wave overlaps the maximum of the other and the two cancel out. Which of these two cases occurs depends on the path difference (*B*) between the two antennas, No. 1 and No. 2. As a source is tracked across the sky, the two signals will alternately reinforce and then interfere as the path difference systematically changes. The combined signals produce *interference fringe* patterns. A controlled electronic delay is placed in the electrical path from antenna No. 1 to the central control, which simulates the delay associated with the path difference, *B*. This assures that the radio source is simultaneously observed by the two (or more) antennas.

The resolution of an interferometer is set by the distance between the two dishes: the farther apart they are, the greater the resolution. There are technological limits on how far apart dishes may be placed and still form a working interferometer. The key issue is that the two radio signals must add together precisely; if one signal should suffer a spurious delay on its way to the central computer, fringes will not be produced. The two radio signals have to be compared at the correct instant or the interferometer does not work.

Over short distances, or *baselines,* cables or waveguides are used to connect the dishes to the central receiver and computer, but when the dishes are separated by more than a few dozen miles, expansion due to small temperature variations will change the length of the cables and cause unpredictable delays in the signals. Thus fringes will be lost.

Over longer baselines, microwave relays can be used to carry the radio signals to the central computer; this technique has been used in England extensively. Beyond a distance of a few hundred miles this technique also breaks down, because the irregularities in the microwave paths between the relay towers are enough to destroy the fringes. For even longer baselines the signals and a very accurate time standard are recorded on magnetic tape at each antenna. The tapes are later synchronized and played back and the data are searched for fringes. This technique appears to have no limit except the accuracy with which time signals can be generated and compared.

Two or more dishes separated by some distance may simulate the resolution of a much larger radio telescope, but they do not have the same collecting area, hence sensitivity, as a single large dish. Fortunately there are plenty of radio sources that today's interferometers and low-noise receivers can observe effectively.

## Very Long Baseline Interferometry

The Very Long Baseline Interferometer (VLBI) system now in operation allows the radio signals from radio dishes separated by thousands of miles to be combined. Joint experiments between the U.S., Canada, England, Sweden, the USSR, Australia, South Africa, and Germany have produced fringes for the tiny cores of quasars and radio galaxies, which have revealed structures as small as 0.0004 arcseconds. This is a resolution 2500 times higher than can be achieved with optical telescopes, and even the Hubble Space Telescope, above the earth's atmosphere, will not do much better than 0.05 seconds of arc.

The U.S.–Canadian VLBI Network involves about a dozen radio observatories which engage in brief two- or three-day observing runs which generate hundreds of 2-inch-wide magnetic tapes containing enormous amounts of data which then take months to process in the central facility designed to handle the tapes.

In some of the radio source maps already shown (for example, in Figures 3.6 and 4.5) the results of VLBI observations were included. Figure 17.3 shows a zoom sequence on the radio galaxy 3C 120, in which VLBI observations were used to produce the contour maps in the lower two frames. Yet the VLBI maps are still relatively crude compared to what will be possible in the next decade (see Chapter 18).

The VLBI technique has been used extensively in observing interstellar masers and is useful for geophysical measurements. Fringes from a radio interferometer can only be generated provided the location of, and hence the separation between, the dishes is precisely known. A consequence of very extensive VLBI observations of many quasars is the ability to position the radio telescopes to an accuracy of a few centimeters. This originally threatened to be a problem when the first Soviet–USA joint experiments were done. At the time, the U.S. radio astronomers did not know precisely where the Crimean radio telescope was located. They had its geographical coordinates, according to astronomy tables, and they knew it was on the shore of the Black Sea, but when the location was plotted on the only map they had of the Black Sea coastline, a World War II German chart, they discovered that the radio telescope was miles out in the Black Sea! To obtain fringes in that experiment a lot of computerized searching of the data had to be done. The experiment was successful, fringes were found, and both countries now know the exact location of their adversary's territory to an accuracy of a few centimeters!

Recent worldwide VLBI observations have revealed that Sweden and Massachusetts are sliding away from each other at 1.7 cm/yr, a number which fits well with continental drift theory. This is not enough to cause concern that transatlantic travel costs will escalate out of control as the travel distance increases with time. Soon VLBI data may also reveal how much slippage there is along the San Andreas fault, a matter of more practical concern.

The VLBI system uses portable receivers and atomic clocks. So as not to lose their synchronization, the clocks have to be kept running at all times, even while being transported to distant radio observatories in remote Scandinavian

FIGURE 17.3. A series of contour maps of the radio source 3C 120, a distant radio galaxy, showing the effect of increasing resolution in revealing ever more detailed structure in the source. The two lower frames are based on data collected with the Very Long Baseline Interferometer operating over intercontinental distances. (NRAO. Observers— R. C. Walker, J. M. Benson, and S. C. Unwin.)

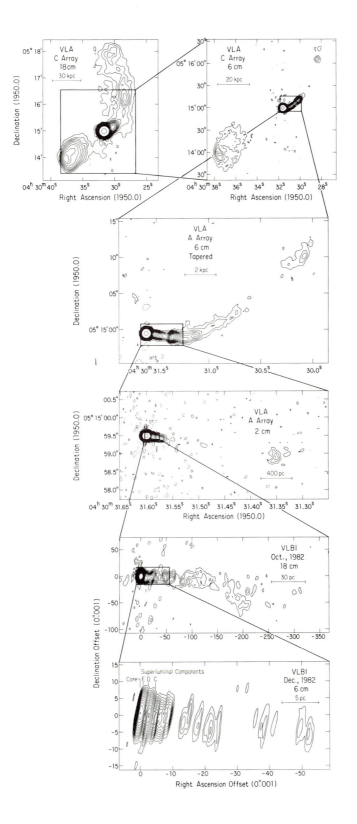

227

or Soviet sites. Radio astronomers, after nursing such clocks in airplane seats across the Atlantic, have been observed to frantically rush off planes in foreign airports to plead with customs agents to allow them to quickly rejuvenate their clock batteries at the nearest power socket. Sometimes they were too late and had to return home, reset the clock, and start the journey all over again.

## Aperture Synthesis

A very beautiful variation of radio interferometry was developed at Cambridge by the radio astronomers led by Sir Martin Ryle. In this technique, known as *aperture synthesis,* the aperture, or area, of a very large dish is synthesized by a few small dishes combined with very powerful computer programs.

Picture two 10-meter dishes located on a football field. If you collect the radio signals from each of these dishes and then move them to every point on the field and combine all the data it is possible to process these data in a computer in such a way that it appears that you have simulated (or synthesized) a radio telescope the size of the entire football field. Admittedly, the process is a slow one, because it takes time to move the two dishes over the entire field. But what Ryle and his team realized was that, as seen from the radio source, any two radio telescopes will appear to rotate around each other during the day due to the rotation of the earth. This is easiest to picture if you imagine one dish placed at the north pole and the other some distance away. During a day the second dish will describe a circle around the one at the pole. This is true of any pair of dishes on the surface of the earth. They will generally describe elliptical paths around each other. After moving one dish relative to the other, a second ellipse will be described, and so on. Further moves of the dishes with respect to each other allow the synthesis technique to be used to fill in patterns which essentially define a much larger-aperture telescope. Enormous apertures can be synthesized in this way. Although the instantaneous collecting area of the larger aperture is not available, the aperture synthesis method samples each part of the large area for a brief period of time, and this sampling simulates the resolution capability of the larger aperture.

To make up for lack of collecting area, astronomers have another trick up their sleeves. Observations are repeated over and over again, which is the equivalent of taking a very long time exposure, and this allows them to detect fainter radio signals.

In practice, aperture synthesis involves more than just two dishes, and the more dishes used, the faster the synthesis proceeds. It also requires very sophisticated computer programs, as well as computers capable of handling vast amounts of data, to make the system work. Modern technological developments, already tested, will soon allow this synthesis technique to be applied to a worldwide radio telescope (Chapter 18).

**FIGURE 17.4.** Two of the 27 25-meter-diameter dishes that make up the Very Large Array near Socorro, New Mexico. Each dish can be lifted off its support pads and carried by rail to another location up to 10 miles away where it is reconnected to function together with the other 26 dishes. (National Radio Astronomy Observatory.)

## The Very Large Array

The world's largest-aperture synthesis telescope is the Very Large Array (VLA), 50 miles west of Socorro, in New Mexico, one of the National Radio Astronomy Observatory's collection of beautiful radio telescopes. The VLA has been used to make many of the radiographs shown in this book.

Twenty-seven individual 25-meter-diameter radio antennas (Figure 17.4) are located along railroad tracks which are laid out in a Y-shape, each arm of which is 21 km long. To completely synthesize the largest possible aperture obtainable by the VLA the individual antennas have to be moved to different locations along the railtracks every few months. A move takes from 4 to 10 days with specially designed transporters carrying the 25-meter dishes to their new stations. Then a series of observations are made before the next change occurs. Later, all the observations from a number of different programs are gathered together and processed in order to synthesize the large aperture.

The VLA is equipped with receivers at 1.3, 2.0, 6.0, and 20 cm wavelengths, which are capable of achieving maximum resolutions of 0.05, 0.08, 0.25, and

0.8 arcseconds, respectively. Radio astronomers have tended to aim for about one-arcsecond resolution, which requires less complete data collection.

Radio waves from space are converted into electrical signals at each antenna and sent to a central control room, where they are collected and later analyzed to make radiographs. Interference fringes are obtained by multiplying together the signals from 351 possible pairs of antennas with the multiplications carried out many times a second. The relationship between the fringes and the radio source structure is a relatively complex one and is derived through an enormous amount of computing. In essence, each small element of structure in the radio source produces a slightly different fringe pattern in each pair of antennas. There is a straightforward mathematical relationship between the location of the source structure on the sky and the nature of the fringes produced. In addition, as the VLA tracks the object across the heavens, the fringe patterns produced by each part of the radio source slowly change. After several hours of data taking, enough information is available to derive a picture of the radio source structure. However, recent advances in programming techniques allow the VLA to be used more rapidly than originally planned. In its "snapshot mode," a useful image of a radio source, but lacking the degree of detail shown in Figure 8.6, for example, can be obtained in only five minutes of observing.

A major breakthrough in the production of radiograph images was associated with the way the programs remove the effect of what are known as *sidelobes*. All radio telescope systems suffer from sidelobe problems, which means that they are weakly sensitive to radio waves coming in from other directions than just the direction in which the telescope is pointing. This is a fundamental limitation of the actual engineering of antennas, whether small Yagi antennas for TV reception or giant radio dishes for radio astronomy. The most serious problems are usually associated with a strong radio source being in a sidelobe well away from the part of the sky being surveyed. Such a source could produce a significant signal which might be confused with a radio source being studied. The way sidelobes are spread around the beam of the radio telescope is usually well known so that their effect can, in principle, be minimized. This is now done very successfully with aperture synthesis telescopes.

An elegant computer program called "CLEAN" is used to remove the effect of sidelobes, and Figure 17.5 shows what this does to the radio image of Cygnus A. The top image shows the relatively crude data, with the sidelobes the cause of the ring patterns. After *cleaning* the image, the middle frame is produced. However, another dramatic technique, only exploited recently, allows the image to be further sharpened.

As was mentioned before, unpredictable errors, or phase delays, in the path from one telescope to the central control room can cause interference fringes to be lost. These phase errors can occur in the cables or in the atmosphere over the individual antennas. This becomes severe for the VLA, where the cloud cover over one antenna may be quite different from that over the other end of the array, 30 or more km away. A technique for taking this into account and removing the phase errors was invented in the 1950s, but it was forgotten until a few years ago when the reports describing the technique were resurrected.

The technique makes use of the fact that there are 351 pairs of radio telescopes involved in looking at the identical radio source structure. But there are only 27 individual antennas, each suffering from its own, unknown, phase errors. As mathematicians know, if you have 27 unknown quantities you require at least 27 equations to solve for these quantities. The VLA provides a continual stream of 351 equations, which are used to solve for the 27 individual phase errors to very great accuracy. The VLA radiographs produced over the last few years are significantly better than those that could be done as recently as 1982.

The third frame in Figure 17.5 shows what the image of Cyg A looks like after this *self-calibration* procedure has been applied. The image has become so clear that the faint radio jet is revealed. This jet was hardly visible in the central frame and completely invisible in the first frame, yet all three images in Figure 17.5 are derived from precisely the same data.

The production of radiographs is now more accurate than it ever was and requires an enormous amount of additional computing time. The computers provided with the original VLA were not planned with this sophisticated data-processing step in mind. To make the very best radiographs of the larger and brighter radio sources, the VLA now requires a supercomputer to handle the data. Such a computer was not available at the observatory and therefore the first two radiographs which exploit the technique most fully, those of the Cassiopeia A and Cygnus A images (Figures 8.6 and 3.5), were done on a Cray computer leased from Digital Productions, a Hollywood graphics studio. For the first time the functioning of a radio telescope has been significantly improved through the exploitation of new computing techniques, developed well after the telescope was built.

The Cas A image is based on 40 hours of VLA observations. Had these data been processed to the point of making the radiograph in the normal way on the computers available at the VLA, it would have taken 100 times longer. The Cray machine spent only 20 minutes of processing time on the task, although 10 hours were used just to read tapes in and out of the machine. The supercomputer will likely become an important tool for future radio astronomy data reduction and image production tasks.

## MERLIN

The British have built a very large array operating at 151 MHz. Eight dishes connected together in the Multi-Element Radio Linked Interferometer Network (MERLIN) are remotely controlled by microwave link from Jodrell Bank, south of Manchester, and are spread over the south and central parts of England. Figure 17.6 is a radiograph of the radio galaxy 3C 330 produced by MERLIN. This radio source was known to be a double at higher frequencies, but the bridge in low-frequency emission which joins the various components had never been seen before. More low-frequency data will add significantly to our knowledge of the physics of radio sources.

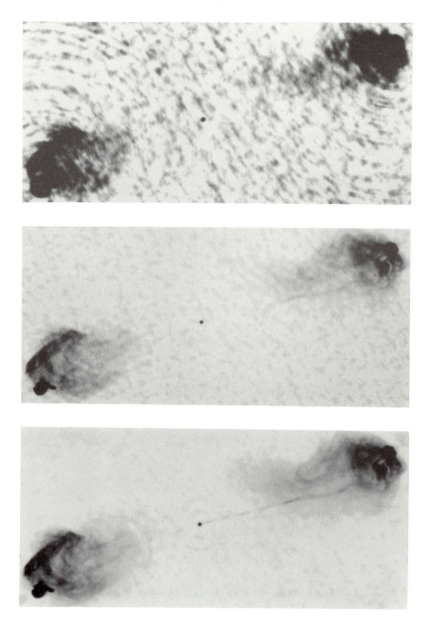

**FIGURE 17.5.** Three images of the radio galaxy Cygnus A which represent three
stages in modern data processing of VLA data. Cygnus A is a radio double with the
main components separated by about 2 minutes of arc. The upper frame represents the
raw data processed to reveal the presence of the radio source, while systematic errors
(sidelobes) introduce the organized patterns of emission which seem to circle around
the two sources. The central frame is the so-called deconvolved image, which represents
the smoothest image that agrees with the original data. This is only one of several
maps which could so agree, although the differences between them would be rather

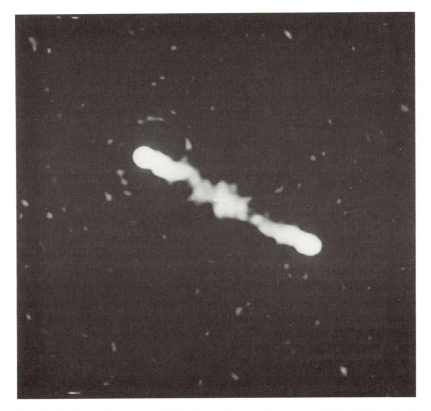

**FIGURE 17.6.** Radiograph of 3C 330 made at a frequency of 151 MHz with the MERLIN system; resolution, 3 arcseconds. The size of the radio jets from one end to the other is 68 arcseconds. The radio galaxy responsible for these jets is located at the position of the white cross in the center. (P. Thomason, Nuffield Radio Astronomy Laboratories.)

small. Recent dramatic improvement in image-processing techniques results from using images such as the central one as the standard and then determining how small errors in each antenna, or along the path through the atmosphere to the antenna, distort the signal. This process, known as self-calibration, leads to the lower image, which has greater sensitivity and allowed the faint radio jet stretching from the central radio source into the northern lobe to be discovered. (NRAO. Observers—P. A. G. Scheuer, R. A. Laing and R. A. Perley.)

# Millimeter Radio Astronomy

One of the most important, relatively unexplored parts of the radio spectrum is in the millimeter-wavelength region. This part of the spectrum, bordering on the infrared, is one of the most heavily populated with spectral lines from complex interstellar molecules. The study of these lines in star-forming regions is likely to lead to an enormous increase in our understanding of the processes of molecule formation and perhaps even of the origin of life itself. In recent years, the world over, radio astronomers have constructed millimeter-wave systems. Several have recently been brought into operation and we will mention a few of them.

## The Caltech Millimeter-Wave Interferometer

Situated in scenic Owens Valley, California, the Caltech Radio Observatory boasted for a long time one of the larger vintage radio telescopes, a 40-meter-diameter dish, as well as a simple interferometer. A new interferometer (Figure 17.7), capable of operating down to millimeter wavelengths, the band particularly important for interstellar molecule studies, was recently brought into operation.

**FIGURE 17.7.** The Caltech millimeter-wave radio interferometer in Owens Valley, California. Three 10.4-meter high-precision telescopes are situated on railroad tracks 400 meters long and operate together to produce high-resolution maps of molecular clouds and other radio sources. (California Institute of Technology.)

This interferometer consists of three 10.4-meter-diameter dishes of extremely high accuracy, placed along railroad tracks extending 300 meters.

## Japan's Nobeyama Radio Observatory

After years of maintaining a relatively low profile in the radio astronomical world, the Japanese finally entered the picture with a beautiful new radio telescope and associated radio receivers. In 1983, at the Nobeyama Radio Observatory, they completed the world's largest millimeter-wavelength dish, 45 meters in diameter, shown in Figure 17.8. An example of a radiograph made with the 45-meter radio telescope showing the radio image of the Rosetta nebula, observed at a wavelength of 3 cm, resolution 2.6 arcminutes, is shown in Figure 17.9. (Figure 6.5 is another example of a radiograph made with this radio telescope.)

The Nobeyama Observatory is also equipped with an aperture synthesis telescope consisting of five 10-meter dishes operating at 1.35 cm and 2.6 mm. The small dishes can be located along railroad tracks 500 meters long, and the interferometer will be capable of 5- and 1-arcsecond resolutions at wavelengths of 1.35 cm and 2.6 mm, respectively.

*45-m Radio Telescope*

**FIGURE 17.8.** The 45-meter-diameter millimeter-wave radio telescope at the Nobeyama Radio Observatory in Japan. This dish is the largest millimeter-wave radio telescope in the world and is being used extensively to study spectral lines from interstellar molecules. Smaller, 10-meter-diameter dishes, used together with the large dish as an interferometer, can be seen in the background. (Nobeyama Radio Observatory.)

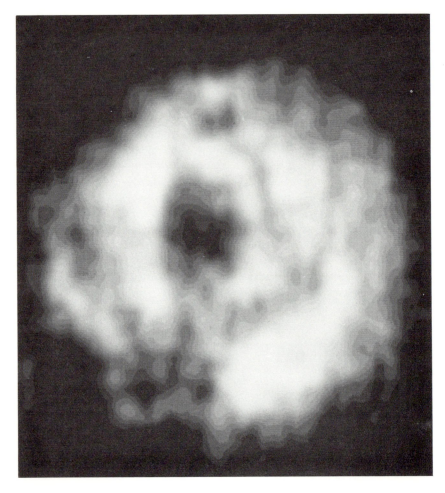

**FIGURE 17.9.** Radiograph of the Rosette Nebula, an HII region, made with the 45-meter telescope shown in Figure 17.8, operating at a wavelength of 3 cm. Vertical size 1°. (Nobeyama Radio Observatory.)

## Europe's Institut de Radio Astronomie Millimetrique (IRAM)

A joint venture of the Centre National de la Recherche Scientifique in France and the Max-Planck-Gesellschaft in West Germany has created IRAM, a radio astronomy center with headquarters in Grenoble, France, specializing in millimeter-wave studies. At the end of 1985 they began observations with their elegant 30-meter telescope located at 2850-m altitude at Pico Veleta, in the Sierra Nevada range of Spain. This radio telescope is the most accurate of its kind in the world and will observe down to 1-mm wavelength, where its sensitivity will be unrivaled. The structure is built upon homology principles, which means that as it moves, the distortions produced by gravity are automatically compen-

**FIGURE 17.10.** Europe entered the millimeter-wave stakes in 1985/1986 with this beautiful antenna located in the mountains of the Sierra Nevada in Spain. This 30-meter-diameter dish is covered by a protective housing which allows its surface and support structure to be temperature-controlled so as to maintain the most accurate surface possible. Operated by IRAM, a joint venture of France and West Germany, the telescope focuses on interstellar molecule studies. (Institut de Radio Astronomie Millimetrique.)

sated, thanks to clever design, so that the telescope is shaped into a series of new paraboloids. Figure 17.10 shows that the telescope has a very different, almost science-fiction-like appearance. This is because the entire structure is enclosed and ventilated so as to keep it at a constant temperature. As a result, the paraboloidal surface remains accurate and remains an efficient reflector of the very short waves.

The 420 surface panels each have a manufacturing accuracy of 25 microns (a micron is one-thousandth of a millimeter) and were adjusted to within 100 microns with respect to each other to form a parabolic surface. Although the structure weighs 800 tons, it is so well balanced that only small motors are needed to steer it.

IRAM also operates a millimeter-wave interferometer consisting of three 15-meter movable antennas located on tracks several hundred meters long. It is situated on the Plateau de Bure in the Hautes-Alpes, about 90 km south of Grenoble. The individual telescopes are unique in that they are built of carbon-fiber-reinforced plastic, stronger than steel, and also use homology principles to control the shape of the surfaces. All three dishes should be in operation by the end of 1987.

# 18

## The Future

### The Very Long Baseline Array

In 1985 work began on the $180,000,000 Very Long Baseline Array (VLBA) project, which will be a continent-sized radio telescope capable of enormously high resolution. Construction responsibility is being shared by the National Radio Astronomy Observatory and Haystack Observatory, with funding by the National Science Foundation. The VLBA should be completed by 1992 and will consist of ten dedicated and automated 25-meter-diameter dishes distributed from Hawaii to St. Croix, eight of them in the continental USA, and located so as to optimize image quality. Each telescope will be independently controlled via telephone lines from the control center in Socorro, New Mexico. Data will be recorded on high-density tapes capable of storing $10^7$ bits of information on one square inch of tape. Processing of the recorded data from all the telescopes in a central computer (capable of about $10^{12}$ multiplications per second) will allow the synthesis of a single radio antenna 8000 km in diameter, physically the largest dedicated telescope on earth. The VLBA will be able to produce radiographs with an angular resolution of two-tenths of one-thousandth of an arcsecond (0.2 milliarcseconds). This may be compared with one arcsecond for the typical radiographs shown in this book. The VLBA will have the potential to see planetary systems as far away as 15,000 light-years.

Because the VLBA will measure radio source structures and positions to a fraction of a milliarcsecond, it will also allow accurate determination of continental drift, as well as the earth's rate of rotation and the wobble of its axis. Even the effect of wind on the rotation of the earth will be detectable. Primarily, though, the VLBA will be sensitive to motion and fine detail within the hearts of the most distant radio galaxies and quasars. In addition to producing radiographs of the nuclei of these objects, it should prove possible to make actual motion pictures of the explosive events feeding the radio jets.

Direct measurement of the distance to objects throughout our galaxy as well as in nearby galaxies, something which is extremely difficult to do by other means, will also be possible. This is a consequence of the extremely high-resolution capability, which will show movement within supernova shells in other galaxies. When such data are compared with Doppler shift observations of the light from a supernova remnant, the distance to the object will be closely estimated. These data will allow the size and age of the universe to be more accurately determined.

The VLBA will probe deeply into gaseous clouds, where stars are being born. Observations of radio signals from molecules in cocoons surrounding the youngest stars should begin to unveil intimate details of the mysterious process of stellar birth.

Antennas in other nations can be linked to the VLBA to form the largest radio telescope that can be spread over the surface of the earth.

# Quasat

After the VLBA is operational, radio astronomers will run out of space! For the first time in history we will have to leave the surface of the earth because our planet is not large enough to satisfy a particular experimental technique any further! The VLBA will simulate a single antenna 8000 kilometers (5000 miles) across. The diameter of our planet is only slightly more than 11,000 km and links with the European VLBI Network will allow resolutions corresponding to a radio telescope with this diameter to be obtained. Beyond that the ground-based interferometer can grow no larger.

The most interesting and dramatic phenomena occurring at the nuclei of galaxies, around the black holes believed to exist there, occur on scales so small that it may require radio telescopes with diameters of many tens of thousands of kilometers to produce the most important images. While the VLBA will be capable of seeing radio source structures with far greater clarity than the VLA, it is expected that more valuable information will be obtained if radio astronomers can observe even smaller-scale structures inside radio galaxies and quasars and in star-forming regions. This can be done by making the VLBA larger by adding just one more antenna, but it has to be located in space.

An exciting plan, now being discussed at the international level, involves the launching of a satellite-borne radio telescope, to be named *Quasat*, which can be used in conjunction with the VLBA and the European VLBI Network (VLN).

Source structure details as small as 70 microarcseconds at 1.3-cm wavelength would be detectable. At the distance of 3C 273 ($1.5 \times 10^9$ light-years) that corresponds to a size of half a light-year. In a radio galaxy such as Centaurus A, $3.3 \times 10^7$ light-years away, details as small as 3 light-days across would be distinguishable.

Quasat will be an international venture for several reasons. The project may be too large for one nation. European, Soviet, American, and Australian radio astronomers are all very interested in this dramatic extension of radio telescope power and want to be involved. Observation time with a radio antenna in space, moving in an extended elliptical orbit (which changes the baseline length from moment to moment), can be easily shared by radio astronomers on different continents.

## The Australia Telescope

Radio astronomical effort in the southern hemisphere has long been focused in Australia and now that nation's government has expressed a further commitment to maintaining Australia's leadership by authorizing the funds for a radio telescope which will be something of an equivalent of both the VLA and the VLBA. The Australia Telescope (AT) is scheduled to begin operation in 1988, the year of Australia's bicentennial. Figure 18.1 is an artist's conception of one part of the telescope. Six 22-meter antennas are to be lined up in what is called a compact array, which will be located at Culgoora near Narrabri. Five of these dishes will be located along a 3-km railroad track so that they can be moved to fill in the baselines required to fully synthesize a single radio telescope 6 km in diameter. This is made possible by locating a sixth antenna 3 km away from the other five.

A long baseline array will be formed by combining this compact array with two other radio antennas, one on Siding Spring Mountain, 100 km to the south

**FIGURE 18.1.** Artist's impression of the Australia Telescope, the large-aperture synthesis system which will open the southern skies to detailed radio astronomical study (see text). (Commonwealth Scientific and Industrial Research Organization.)

of Culgoora, and the other the existing 64-meter Parkes radio telescope 200 km further south. The combination of these eight antennas will synthesize an aperture 300 km in diameter.

The radio signals from each of the six dishes in the compact array will be brought to the central control room by means of optical fiber transmission, and the large array will use tape recorders to collect all the data. In time the system can be expanded to include radio antennas across the width of Australia, so that the equivalent of the VLBA will be created in the southern hemisphere. The diameter of the Australian VLBA would be 3000 km. After these stages are completed the Quasat antenna may be used together with the AT.

## A Proposed Millimeter Array

With the plans for the VLBA and the AT, and the recent completion in Japan and Spain of large millimeter-wave dishes, there are very few directions radio astronomers can move to improve their ground-based radio telescopes to open new windows into the universe. One project being considered is the construction of a large-aperture synthesis array of highly accurate smaller dishes to operate at millimeter wavelengths. An array of 21 10-meter dishes, spread over a baseline of 35 km, combined with a multi-telescope array of many smaller 3- or 4-meter dishes, all mounted in a single structure to provide a large collecting area, is being discussed.

## How Much Longer Will Radio Astronomy Last?

This question is not asked lightly. Radio astronomers observe in the radio frequency part of the electromagnetic spectrum, but there are many people who would like to use those same radio bands for other purposes. Some of the more obvious groups (or services) with an interest are the communications industry, the military, and NASA.

Countless radio transmitters already operate on our planet, some for satellite communications, some for military communications, some for TV and FM or broadcast radio. There are many other users such as radiotelephones, taxis, the National Guard, commercial airliners, private planes, navigation satellites, military reconnaissance satellites, and even your local radio-dispatched plumber. All want to use parts of the radio band. However, any radio transmitter is a potential threat to radio astronomy. The problem of unwanted radio interference is already acute.

A radio signal leaking from a communications satellite can destroy hours of astronomical observation. To the radio astronomer such a signal appears as a flashbulb might to an optical astronomer trying to take a photograph of a distant galaxy. The air waves, a natural resource, are increasingly being commandeered by those who wish to exploit the radio frequency spectrum for commercial or military purposes. Radio astronomers feel this takeover very keenly, and although they have a voice in the World Administrative Radio Conference, which recommends on matters regarding the use of the radio frequency spectrum by various services, radio astronomers suffer from a weakness in their lobbying power. This is due, in part, to their having neither commercial, military, political, nor national security interests behind them. Inexorably the blinds on the radio window to the universe are being drawn.

At the root of the problem is the radio astronomers' desire to keep the radio spectrum as quiet as possible. All they want to do is listen to faint cosmic whispers. Other services want to use those same bands for what each service regards as very important business—whether it be for dispatching taxi cabs, detecting nuclear explosions, or beaming satellite programs all over the planet. The irony is that while international agreement has created *protected bands* for the various services, such "protection from services in other bands shall be afforded the radio astronomy service only to the extent that such services are protected from each other." However, most other services do not care how much spurious radio energy is leaking from one radio band into another, because such leakage is usually well below the levels of interest to them. However, this leakage, due to poor design of transmitters, threatens to destroy radio astronomy observations. At issue is the fact that while a given service does make some effort to prevent excess leakage of unwanted signals into other bands, this concern is relative. Radio telescopes, capable of picking up faint signals from quasars 15,000,000,000 light-years away, are completely swamped by even otherwise faint satellite signals in antenna sidelobes at levels which satellite transmitter designers do not dream of as being of interest to anyone.

To cut down on radio frequency pollution, special filters can be built into the transmitters, but they add to the expense.

The situation has become so serious that observations of (redshifted) OH in other galaxies, for example, which are ideally done with the Arecibo telescope or the VLA, have become virtually impossible because of global satellite systems launched by both the USA (Geostar) and the USSR (Glonass). Some nations are taking the disregard for radio astronomy interests further. A joint USSR/French spaceflight to Venus communicates with the earth at exactly one of the frequencies of the OH molecule, which lies in a partially protected radio astronomy band.

The U.S. Air Force NAVSTAR satellites, of which 18 are planned, will blanket the planet with radio transmission in the 1370–1390-MHz band, within which redshifted hydrogen gas in distant galaxies shines faintly. More importantly, the lack of design concern for leakage from these satellites into adjoining bands, due to ineffective filters on board the satellites, may destroy 21-cm observations of hydrogen in our galaxy. This would be disastrous to astronomy. The consequences of such behavior are not very different from what might be experienced if the Air Force were to launch 18 powerful searchlights to shine down onto the USA to see what is going on here at night. Eighteen searchlights in space would blind us and prevent us from seeing the stars.

The radio astronomy community has unusual needs which those interested only in communications do not see as a priority. We hope that some of those involved in communications may read this book and see the radiographs presented here. Then they may care to better protect those parts of the radio spectrum so critical to radio astronomy. The very real threat is that the wonders of the invisible universe, now being so dramatically revealed, may soon be rendered invisible because of radio pollution generated by the technological nations on our planet.

## Radio Astronomy and Imagination

And so we have come to the end of our story. The invisible universe has been revealed in images which even a decade ago would have been impossible to obtain. Although this book contains only static pictures, each radiograph is a snapshot of an object in a state of continual upheaval. The motion, the chaos, and the violence found in the invisible universe are recognized only when you wrap your imagination around the images. Look again at Cassiopeia A (Figure 8.6) or Cygnus A (Figure 3.5) and imagine how movement plays a role. Do not hesitate, because your imagination is as valid as the next person's in trying to visualize this.

To fully appreciate the new radio discoveries requires the continual involvement of your imagination. Exploration of the cosmos becomes an adventure when it takes place inside the mind. The explosion of a quasar is not witnessed in space somewhere, but in your imagination. Radio waves from space do not arrive prepackaged in the form of radiographs. Human imagination was required to design and operate the appropriate radio telescopes, receivers, computer programs, and photographic equipment to present the discoveries in forms which allow us all to relate to them. Many people have brought a lot of creative thinking to bear on the data in order to form the concepts and pictures presented in this book. For us to relate to what has been revealed about the nature of the invisible universe we have to involve our imaginations.

The dynamical aspects of astronomy are revealed not by what is seen at the far end of the telescope, but by what is experienced at this end. This is where the real excitement is to be found. The human mind has discovered that distant phenomena in the seemingly changeless universe reveal continual upheaval and evolutionary processes in action. Thanks to the workings of the human mind, aided by physics, mathematics, and computers, astronomers can simulate cosmic phenomena and so speed up time. Such simulations allow us to recognize how evolution, change, and catastrophic events shape distant gas clouds, galaxies, and quasars.

The choreography of the physical universe contrasts ponderous motion and electric action. The discovery and communication of the cosmic dance have hardly begun. It is hoped that you will have sensed a little of its rhythms during this exploration of the invisible universe as revealed by radio astronomy.

# Appendix

## 1. Atoms, Ions, Molecules, and Cosmic Rays

An *atom* is a basic unit of matter. It consists of three fundamental particles: electrons, which carry a negative electric charge; protons, which carry an equal but positive electric charge; and uncharged neutrons. A typical atom consists of a nucleus of protons and neutrons with electrons orbiting it. The simplest atom, hydrogen, has one proton at its nucleus and one electron orbiting it. The next atom on the scale of complexity (and mass) is helium, consisting of a nucleus of two protons and two neutrons with two electrons orbiting it. Carbon, one of the basic building blocks of life, contains six protons, six neutrons, and six electrons. When an electron is removed from an atom, the atom is said to become *ionized*.

*Molecules* are formed by linking atoms, with their electrons acting as a glue to hold them together. The air we breathe consists mostly of oxygen molecules (pairs of oxygen atoms) and nitrogen molecules (pairs of nitrogen atoms). Molecules can sometimes be extraordinarily complex and may consist of thousands of atoms strung together in great chains, for example, DNA in living cells.

A *cosmic ray* is an electron or proton hurtling through space at close to the speed of light. When a particle travels close to the speed of light, it exhibits properties predicted by Einstein's Special Theory of Relativity, and so is said to be *relativistic*. Relativistic electrons are important in astronomy because they produce radio waves in a wide variety of astronomical situations, many of which are described in this book. Some cosmic-ray particles in our galaxy originate in dying stars, others in explosions on the surface of the sun.

## 2. Stars, Planets, Galaxies, and the Milky Way

We live on a *planet* which orbits a *star* called the sun. Stars generate their own heat, light, and radio signals. Our sun is one of 250,000,000,000 stars in our galaxy which, seen from a distant perspective, probably looks like the galaxy M74 in the constellation Pisces (Figure A.1). When we look out along

**FIGURE A.1.** The spiral galaxy M74, also known as NGC 628, in the constellation of Pisces. From a great distance our galaxy may look very much like M74, in which stars and dark interstellar material (dust) lie in spiral streamers known as spiral arms. The isolated points of light in the photograph are foreground stars in our galaxy. (National Optical Astronomy Observatories.)

the disk of the galaxy we see so many stars that their combined light appears as the diffuse band of light which crosses our night skies in the northern summer. That band is known as the Milky Way, a name also given to our galaxy, which is about 100,000 light-years (see Section 3 below) across and 1000 light-years thick.

## 3. Astronomical Distances—Looking Back in Time

Astronomers cannot avoid seeing back in time when they look out into space. Light and radio waves traveling at the speed of light come to us from great distances and have been on their journeys for long periods of time. Astronomers are doomed to peering into the past! They are used to this concept and to them it is second nature to think of great distances in terms of vast spans of time. Even the most often used unit of distance, the light-year, is based on the distance a light beam can travel in a year. The term *light-year* allows us to encompass a huge distance ($6 \times 10^{12}$ miles) in two words. Astronomers usually use the *parsec* as the unit of distance, a parsec being about 3.26 light-years.

# 4. Notation

One million can be written as $10^6$, i.e., a 1 followed by six zeros = 1,000,000. For numbers smaller than unity the notation is similar, e.g., $10^{-2} = 0.01$, or one-hundredth. A light-year is about $6 \times 10^{12}$ miles. More useful to remember is that a light-year is about $10^{18}$ cm.

This superscript notation is also used in another way. If a gas has a density of a million atoms per cubic centimeter, astronomers write it as $10^6$ cm$^{-3}$.

A common tradition in astronomy is to refer to the mass of astronomical objects in terms of the mass of the sun. One *solar mass* is about $2 \times 10^{33}$ grams, a number far too great for us to comprehend. It is easier to think in terms of solar masses, which let us relate astronomical mass to something closer to home.

# 5. Beyond the Galaxy

Surrounding our galaxy, within a distance of a few million light-years, about 30 other galaxies drift together in what is known as the *Local Group*. Beyond the boundaries of the Local Group there exist several hundred billion other galaxies, most of them gathered in *clusters,* some containing a few dozen members, others thousands of galaxies locked together by their mutual gravitational attraction. The farthest objects yet observed may be 10–15 billion light-years distant.

# 6. How Radio Telescopes Work

For hundreds of years, ever since Galileo, in 1609 A.D., first used an optical telescope to study the moon, stars, and planets, astronomers have used glass lenses or mirrors to gather and concentrate light from distant stars and galaxies. The light is then passed through more lenses to bring it to a focus on a photographic plate or on an electronic detector in the best of modern telescopes.

A radio telescope is similar to an optical telescope, but it reflects radio waves off a metal surface instead of a glass mirror. The larger the reflecting surface, the greater the amount of energy gathered and the fainter the radio signals that can be sensed. The largest single "dish" radio telescope that has been built is near Arecibo, Puerto Rico, and has a diameter of 1000 feet (see Figure 7.1).

Figure A.2 shows the basic radio telescope. A radio wave from space is

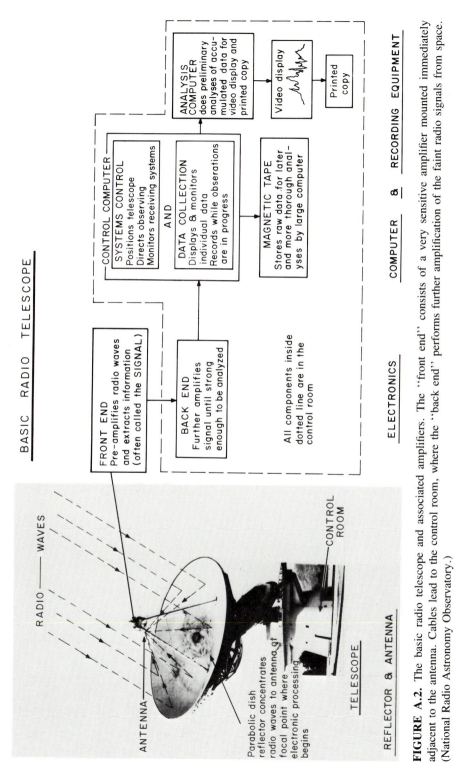

**FIGURE A.2.** The basic radio telescope and associated amplifiers. The "front end" consists of a very sensitive amplifier mounted immediately adjacent to the antenna. Cables lead to the control room, where the "back end" performs further amplification of the faint radio signals from space. (National Radio Astronomy Observatory.)

reflected off the parabolically shaped surface to a focus at which is placed a small antenna, which may look similar to a conventional TV or FM antenna. There the concentrated radio signals are converted into minute electrical currents in amplifiers connected to the antenna. This is known as the "front end" of the receiver. These currents are then sent to the control room where they are amplified a million or more times in the "back end" of the receiver before being processed in a computer or displayed in such a way that the radio astronomer can "see" what the data indicate.

## 7. The Electromagnetic Spectrum

Figure A.3 shows the full range of the electromagnetic spectrum. Radio waves are at the long-wavelength end (hundreds of meters), followed by infrared (IR) radiation, commonly experienced as heat, then light waves and ultraviolet (UV) radiation, sometimes called "blacklight." (UV causes sunburn, and in large doses is extremely harmful to living organisms.) Next are X-rays, with wavelengths so short that they literally wriggle between atoms and so can penetrate our bodies. Finally, at the shortest-wavelength (less than $1 \times 10^{-8}$ cm) end of the spectrum are the gamma rays.

Today astronomy is far more than peering through optical telescopes. There is radio astronomy, X-ray astronomy, IR astronomy, UV astronomy, and even gamma-ray astronomy, each with its special types of telescopes and each sensitive to an invisible universe which is different from the one we can see with our eyes alone.

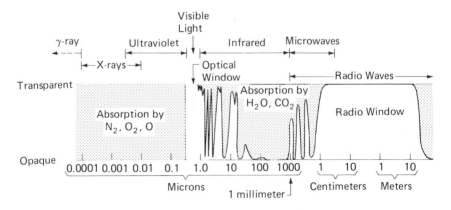

**FIGURE A.3.** The extent of the electromagnetic spectrum, indicating the location of "windows" into space. Radio waves between about 20 m and 2 cm pass unimpeded through the atmosphere, as do light waves. However, atmospheric atoms and molecules such as nitrogen, oxygen, ozone, water, and carbon dioxide absorb radiation in other parts of the spectrum.

# 8. Atmospheric Windows

The earth is surrounded by an atmosphere which cuts off most of the UV, IR, X-rays, and gamma rays from space. However, the atmosphere is transparent to radio, light, and some infrared waves, providing two windows into space, as indicated in Figure A.3. Radio astronomy can usually be done during the day or night, independent of the weather. However, water vapor ($H_2O$) and carbon dioxide ($CO_2$) in the atmosphere absorb the shortest-wavelength cosmic radio waves. Terrestrial clouds are, in fact, opaque to millimeter waves, while 1-cm radio waves are partially absorbed. Short-wavelength radio astronomy can best be done in dry climates at high altitudes (in order to be above as much atmospheric water vapor as possible), but as soon as clouds appear, observations cease. IR, UV, X-ray, and gamma-ray astronomers generally require either balloon-borne telescopes or satellites in order to make their observations above the earth's protective blanket because the nitrogen and oxygen in the atmosphere absorb these radiations.

# 9. Wavelength and Frequency

In referring to radio waves, the terms *wavelength* and *frequency* are often used and are related in a simple manner. Consider waves crashing into the seashore or lapping the edge of a lake. You may notice that the waves strike the shore at a certain rate, or *frequency*. Along the California coast a typical interval between waves is ten seconds, which indicates a frequency of six waves per minute. The frequency at which the water waves break is related to their speed; the faster they travel, the greater their frequency. Waves also have a characteristic wavelength, the distance between crests.

A radio wave travels at the speed of light, 186,000 miles/second, or 300,000 km/s. (We use the metric system and kilometers per second is abbreviated to km/s.) A radio wave one meter in length has a frequency of 300,000,000 cycles/second. This is written as 300 megahertz (for millions of hertz, the modern unit for frequency, named after the physicist Heinrich Hertz). Megahertz is further abbreviated to MHz. Similarly, a radio wave of 1-centimeter (cm) wavelength has a frequency of 30,000 MHz. Frequency is given by the speed divided by the wavelength.

# 10. Spectral Lines

Light from distant stars, galaxies, and quasars often contains energy at very specific wavelengths. These signals are known as spectral lines. Chapters 9 and 10 describe a little more about how some of them originate. Suffice it to say that spectral lines are usually due to radiation of energy from particular atoms and that each atom has its own characteristic signature of spectral lines which astronomers are trained to recognize.

# 11. The Redshift

The redshift is the name given to the stretching of light waves, or any other electromagnetic waves, produced by movement of the source of the radiation away from the observer. To illustrate this, recall the sound of a jet plane. When it is flying toward you, the sound is high-pitched. As the plane passes and flies off into the distance, the sound moves to a lower and lower pitch. This is known as the Doppler effect, after the Austrian physicist (Christian Doppler) who first studied the phenomenon of the change in frequency of waves from a moving source. When the plane flies away the sound waves it emits are stretched—that is, their wavelength becomes greater. That produces a lower tone. Similarly, when a galaxy or star is receding from the earth its light waves are slightly stretched, which means they shift toward the longer, or red end, of the spectrum—hence a redshift. The opposite effect, produced by the object coming toward you, would produce a shortening of the waves, or a blueshift.

# 12. Astronomical Coordinate Systems

The geographical coordinates of latitude (north–south) and longitude (east–west) are used to locate objects on the surface of our planet. Astronomers use similar angular coordinates to locate objects on the sky. (Note: Angles are measured in degrees (°), minutes ('), and seconds (") of arc. A full circle contains 360 degrees. Each degree consists of 60 arcminutes, and each *arcminute* consists of 60 arcseconds.)

Imagine drawing a line across the heavens which is always directly above the earth's equator. This is known as the *celestial equator*. The angle measured north and south from this celestial equator is called *declination* (directly equivalent to latitude on earth). The North Celestial Pole, located directly over the North Pole of the earth (just about where the Pole Star is found), is at +90° declination.

The astronomical equivalent of terrestrial longitude, the coordinate measured east and west of Greenwich, England, is called *right ascension* and can also be given as an angle around a circle, but is commonly measured as a time— 24 hours span the equator which is a full circle of 360°. Right ascension, in hours, minutes, and seconds of time, is measured east of an agreed-upon zero point known as the First Point in Aries. Although there are technical complications associated with the precise definition of these coordinate systems, suffice it to say that right ascension and declination are the basic astronomical coordinates.

Another system of coordinates is based on the Milky Way. A line defining the central plane of the galaxy, a line which runs along the center of the Milky Way band of stars, is defined as the *galactic equator*. *Galactic latitude* is measured in degrees, minutes, and seconds of arc north or south of this equator, and *galactic longitude* (also in degrees, minutes, and seconds of arc) is the angular distance measured along the galactic equator, using the direction of the center of the galaxy, in the constellation of Sagittarius (see Chapter 6), as the zero point.

## 13. The Brightness of Radio Sources

The strength, or intensity, of radio waves received from a distant radio source is usually given in terms of a unit which is defined by international agreement. Such a unit is the *jansky,* named after Karl Jansky, the discoverer of the radio waves from the Milky Way. A jansky is a measure of the amount of radio energy striking a given area (one square meter) in a specific frequency interval (one Hz): $10^{-26}$ watts per square meter per Hz. Radio astronomers also describe the intensity or strength of a received radio signal in terms of a temperature. The *antenna temperature,* given in terms of degrees Kelvin, is the temperature the universe would be if it were to radiate the same power as is captured by the radio telescope observing that specific source. Luminous radio sources may produce very large antenna temperatures, depending on whether or not they fill the beam of the antenna. For a very small, pointlike radio source, which may be intrinsically bright, a small antenna temperature will be produced because the source covers a small area of sky as compared with the beam of the telescope. (Many diagrams in this book show either antenna temperature or janskies to indicate the intensity of the received radio signals.)

The *luminosity* of a radio source refers to the actual amount of energy it emits whereas the *brightness* of the source is a measure of the power per unit area radiated by the source. For example, a flashlight may appear bright when placed close to one's face, but across a football field it will appear quite faint. Distant stars also appear very faint in the night sky, but if we should move close to a star we would find that it is enormously more luminous than the flashlight. These terms can be used to refer to optical or radio emission from astronomical objects.

For completeness it should be pointed out that the measured brightness of a radio source can be used to infer the amount of energy actually generated at the radio emitter, provided the distance of the source is known.

## 14. Velocities in Radio Astronomy

At various places in the text—in particular, in Chapters 9 and 10—velocities are sometimes mentioned. In radio astronomy the velocities for galactic matter are given in terms of a Doppler shift of the spectral line with respect to a reference known as the *local standard of rest* (lsr), defined by international agreement. The sun actually moves with respect to the lsr, which is representative of the way local stars and gas are moving as a whole through the galaxy. It is useful to refer to velocities of distant gas with respect to this standard because the earth is constantly moving through space and at any given moment the observed Doppler shift of a distant hydrogen cloud, for example, depends on how the earth is moving with respect to it. This convention is different from the one used by optical astronomers who define an object's (star or galaxy) velocity with respect to the sun as the reference point.

# Further Reading

*Astronomy Transformed.* David O. Edge and Michael J. Mulkay. Wiley Interscience, 1976.

*Classics in Radio Astronomy.* W. T. Sullivan III, editor. Reidel, 1982.

*Cosmic Ecology.* G. A. Seielstad. University of California Press, 1983.

*The Early Years of Radio Astronomy.* W. T. Sullivan III, editor. Cambridge University Press, 1984.

*The Evolution of Radio Astronomy.* J. S. Hey. Science History Publications–Neale Watson Academic Publishers, 1973.

*Galactic and Extragalactic Radio Astronomy,* 2nd ed. G. L. Verschuur and K. I. Kellermann, editors. Springer-Verlag, to be published. (For a technical approach.)

*The Invisible Universe: The story of radio astronomy.* Gerrit L. Verschuur. Springer-Verlag, 1974. (For an overview of the subject in the 1960s and early 1970s; out of print.)

*Out of the Zenith: Jodrell Bank 1957–1970.* Bernard Lovell. Harper and Row, 1973.

*Radio Telescopes.* W. N. Christiansen and J. A. Hogbom. Cambridge University Press, 1985.

*Serendipitous Discoveries in Radio Astronomy,* Proceedings of a workshop held at the National Radio Astronomy Observatory, May 1983. K. Kellermann and B. Sheets, editors. Published by the NRAO, Green Bank, WV 24944.

*SETI Workshop.* Report of the SETI workshop held at the National Radio Astronomy Observatory, May 1985. K. Kellermann and B. Sheets, editors. Published by the NRAO, Green Bank, WV 24944.

*Solar Radio Astronomy.* M. Kundu. John Wiley, 1965.

General reading for articles of relevance: These appear regularly in *Mercury* magazine, Astronomical Society of the Pacific; *Scientific American;* and *IAU Symposia,* Reidel Press.

# Index